Scilab
from Theory to Practice

I. Fundamentals

by
Philippe Roux
translated by Perrine Mathieu

Scilab from Theory to Practice - I. Fundamentals
by Philippe Roux, translated by Perrine Mathieu

ISBN (paper) : 978-2-8227-0293-5

Published by Éditions D-BookeR, 5 rue Delouvain, 75019 Paris
www.d-booker.com
contact@d-booker.fr

Original title : Scilab de la théorie à la pratique - 1. Les fondamentaux
Original ISBN : 978-2-8227-0053-5

Scilab good practices validated by : Dr. Claude Gomez (Co-Founder and Advisor of Scilab Enterprises)
Cover design : Marie Van Der Marlière (www.marie-graphiste.com)
Layout : made with Calenco / XSLT developed by NeoDoc (www.neodoc.biz)
Dépôt légal (France) : Mars 2016

Date of publication : 03/2016
Edition : 1
Version : 1

Table of Contents

Preface

by Dr. Claude Gomez

Co-Founder and Advisor of Scilab Enterprises

Welcome to *Scilab: From theory to practice.*

Scilab is a computation and numerical simulation software that has today become a model for open source softwares. It boasts more than 1600 mathematical and graphical functions along with Xcos for simulations of hybrid dynamic systems. This means that, despite the online help, users can often find it difficult to harness all its capabilities without real support. Since the fields it covers span all those of applied mathematics, matrix operations, simulation, signal processing, statistics, controls, etc., one would need a book for each. This is what *Scilab: From theory to practice* offers by gradually tackling each of these fields in order to provide a complete overview of the Scilab software. The software world is punctuated by regular updates and traditional paper supports quickly become obsolete. D-BookeR's innovative approach lets you rely on an online book that is always up to date.[1]

Fundamentals is the volume you need to start with. Indeed, by covering the subject gradually and extensively it aims to help you familiarize yourself with the Scilab language and plots. Compared to the naive programming usually produced by a beginner programmer, it shows how to efficiently use the whole extent of Scilab's programming power and shows how to master the different graphics capabilities, 2D, 3D and animations. Thanks to this knowledge, the user will be ready to efficiently tackle the fields of applied mathematics covered in the other chapters.

Enjoy the book and good Scilab practice!

Paris, 31 mai 2013

[1] All the versions are not yet available in English, Ed.

Foreword

Scilab has greatly changed since 2010, consequently a large portion of the existing documentation is now obsolete. This book is based on the most recent version of Scilab (5.5) and we have put a great deal of care into communicating the best practices relevant to the current software.

Since it was initially produced in a digital format, it contains interactive features (hyperlinks) and multimedia (GIF animations, videos), which depending on your version will be more or less integrated. We have nevertheless made an effort to ensure that each version be as user-friendly as possible and to offer easy access to the supporting multimedia material over the internet.

In the printed version, a QRCode located on the right side of the videos or GIF animations substitute image directs you to the online gallery. These galleries accesses are provided below :

- video gallery [http://d-booker.jo.my/scilab1-videos-en] ;
- GIF animation gallery [http://d-booker.jo.my/scilab1-gif-en] ;

Caution › Some URLs (surrounded by brackets) may be interrupted by a line break. Do not include the -dash symbol when copying the link.

Target public and prerequisites

This book is aimed at an audience of new users as well as at individuals familiar with Scilab who wish to update or expand on their current knowledge. It assumes the reader feels comfortable using a computer and possesses a basic knowledge of what computer programming is. Some technical notions as well as physics or mathematics knowledge may be required in some sections.

If you are a completely new user, start by getting to know the software with the help of Scilab for very beginners [http://d-booker.jo.my/sci-beginners], a document created by Scilab Enterprises.

Source code of examples

The source code for the examples can be downloaded at the book's introduction page on the Editions D-BookeR website [http://d-booker.jo.my/sci-basics-book], by clicking on the COMPLEMENTS tab or directly from this link [http://d-booker.jo.my/sci-basics-examples].

Getting Started

Scilab is a multi-platform numerical computation software that is free, open source and boasts a large user community. This first part aims to help you set up and get a grasp of the Scilab work environment.

1 Preview of Scilab

Scilab is a numerical computation software that is open source, multi-platform and is meant for the theoretician as well as the engineer. It possesses a high level computer language that is well adapted to mathematical notation, as well as complementary modules targeted at scientific applications. These applications include signal and image processing, statistical analysis, mechanical simulations, fluid dynamics, numerical optimization, modeling and simulation of hybrid dynamic systems, etc. To demonstrate its capacities, we showcase an example of a comet in orbit around the Sun, following a trajectory perturbed by the presence of a planet in a stable orbit around the Sun. Despite the complexity of the problem, you can easily simulate such a system with Scilab and obtain a three dimensional animation of the movement of the different bodies. An example of this video can be seen below.

Figure 1.1 : Simulation of a comet's orbit (vidéo)

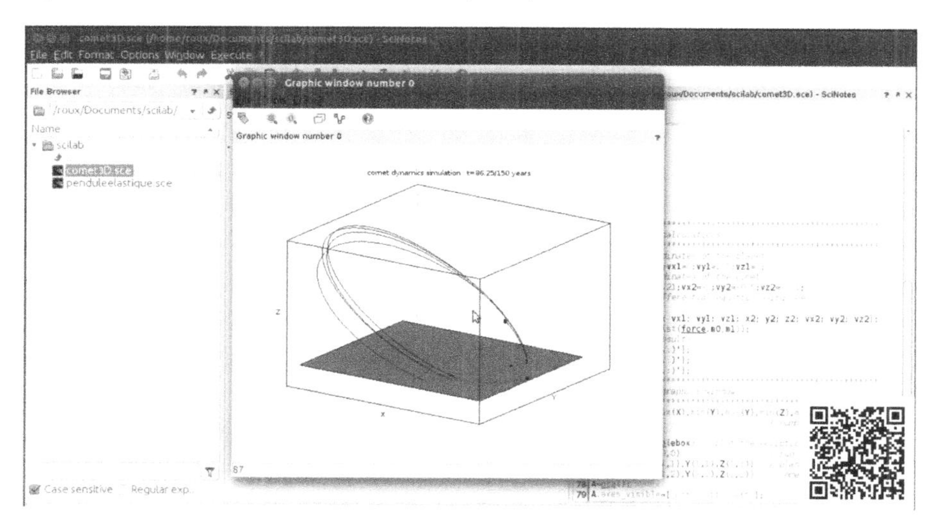

This animation was obtained using a program written in the Scilab language and presented in the following Example 1.1.

Example 1.1 : *Script used to visualize the trajectory of a comet around the Sun*

```
//*********************************************************
// simulation of the perturbed trajectory of a comet
//*********************************************************
//  parameterization of a sphere
function [x,y,z]=sphere(theta,phi)
   A=0.1,B=0.01
   x=A*cos(phi).*cos(theta)
   y=A*cos(phi).*sin(theta)
   z=B*sin(phi)
endfunction
// function to draw a sphere
function plot_sphere(x,y,z)
   phi=[0:0.1:2*3.15];
   theta=[2*3.15:-0.05:0];
   [dx,dy,dz]=eval3dp(sphere,theta,phi);
   surf(x+dx,y+dy,z+dz);
endfunction
// function to plot the z=0 plane
function plot_ecliptic(ebox)
   x=[ebox(1);ebox(2)]
   y=[ebox(3);ebox(4)]
   z=zeros(2,2)
   surf(x,y,z)
endfunction

// functions calculating the gravitational forces
function [u2]=force_g(t,u,mass)
   module=-G*mass*((u(1)^2+u(2)^2+u(3)^2)^(-3/2))
   u2=[module*u(1); module*u(2); module*u(3)]
endfunction

function [du]=force(t,u,mass0,mass1)
   u1=[u(1);u(2);u(3)]
   du1=[u(4);u(5);u(6)]
   u2=[u(7);u(8);u(9)]
   du2=[u(10);u(11);u(12)]
   ddu1=force_g(t,u1,mass0)
   ddu2=force_g(t,u2,mass0)+force_g(t,u2-u1,mass1)
   du=[du1;ddu1;du2;ddu2]
endfunction

// constants
G=0.04;
m0=1000;
m1=1;
dt=0.05;
T=50;
dx=0.5;
dy=0.5;
dz=0.5;
alpha=65;
Beta=150;
```

```
//*******************************************************
// trajectory calculations
//*******************************************************
// initial coordinates of the planet
x1=5;y1=0;z1=0;vx1=0;vy1=2.5;vz1=0;
// initial coordinates of the comet
x2=6;y2=6;z2=0.21;vx2=-2;vy2=-0.5;vz2=-0.1;
//solve the differential equation using ode
t=[0:dt:T];
u0=[x1; y1; z1; vx1; vy1; vz1; x2; y2; z2; vx2; vy2; vz2];
u=ode(u0,0,t,list(force,m0,m1));
// retrieve results
X=[u(1,:)',u(7,:)'];
Y=[u(2,:)',u(8,:)'];
Z=[u(3,:)',u(9,:)'];

//*******************************************************
// launch the graphics window
//*******************************************************
ebox=[min(X),max(X),min(Y),max(Y),min(Z),max(Z)];
N=length(t);                          // number of steps
drawlater()
plot_ecliptic(ebox) // plot the ecliptic plane
plot_sphere(0,0,0)                    // sun
plot_sphere(X(1,1),Y(1,1),Z(1,1))  // planet
plot_sphere(X(1,2),Y(1,2),Z(1,2))  // comet
A=gca();
A.axes_visible=["off" "off" "off"];
A.rotation_angles=[alpha Beta];
A.data_bounds=ebox;
drawnow()

//*******************************************************
// main loop creates the graphical animation
//*******************************************************
for k=1:5:N
   Beta=Beta+k/300;                            //  view angle
   realtimeinit(0.05)                         // unit of time
   drawlater()                   // open the graphical buffer
   clf()                        // erase the graphical buffer
   plot_ecliptic(ebox)        // plot on ecliptic plane
   param3d1(X(1:k,:),Y(1:k,:),...             // display the
   list(Z(1:k,:),[5,2]))                     // trajectories
   plot_sphere(0,0,0)                             // the sun
   plot_sphere(X(k,1),Y(k,1),Z(k,1))          // the planet
   plot_sphere(X(k,2),Y(k,2),Z(k,2))           // the comet
   title('comet dynamics simulation : t='+msprintf(...
   '%2.2f',t(k))+'/'+string(T)+' years')           // title
   xinfo(string(t(k)))                       // display time
   A=gca();                     // resize the graphics window
   A.axes_visible=["off" "off" "off"];
   A.rotation_angles=[alpha Beta];     // rotate pt of vue
   A.data_bounds=ebox;
   drawnow()                    // display graphical buffer
   realtime(k)             // pause to adjust display rate
end
```

This example provides a small preview of Scilab's capabilities pertaining to graphics as well as numerical computation. With a small degree of skill, you will rapidly be able to go beyond this level. For example, you may then create programs that allow the user to interact with Scilab's graphical interface and create his/her own animations such as those shown in the video below.

Figure 1.2 : *Simulation of a pendulum hung from a spring (video)*

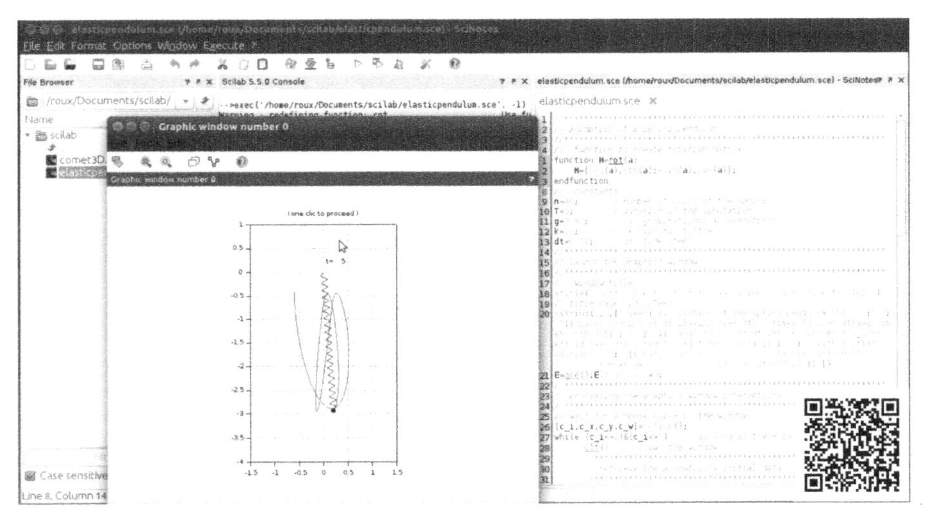

In this example, we simulate the trajectory of a pendulum hung from a spring (of stiffness k), oscillating freely after being released from a given position with no initial velocity. The pendulum's motion is modeled through a system of differential equations involving two variables:

$$\begin{cases} r \times \frac{\mathrm{d}^2 a}{\mathrm{d}t^2} + 2\frac{\mathrm{d}r}{\mathrm{d}t} \times \frac{\mathrm{d}a}{\mathrm{d}t} = g \times \sin(a) \\ \frac{\mathrm{d}^2 r}{\mathrm{d}t^2} - \frac{k}{m}(r - r_0) = r \times \left(\frac{\mathrm{d}a}{\mathrm{d}t}\right)^2 + g \times \cos(a) \end{cases}$$

where:

- a is the angle of the pendulum with respect to the vertical
- r is the length of the spring that constitutes the pendulum

and with the initial conditions: $a(0) = a_0,\ \frac{\mathrm{d}a(0)}{\mathrm{d}t} = 0,\ r(0) = r_0,\ \frac{\mathrm{d}r(0)}{\mathrm{d}t} = 0.$

In order to study a system like this that is very sensitive to initial conditions, it is important to be able to easily modify the initial position and observe the effect of each parameter on the movement. With Scilab, it is possible to create a graphical interface that simplifies the input of parameters and lets the user focus on the analysis of results. With the script shown in Example 1.2, you can modify the initial position of the pendulum with a simple mouse click in the graphics window and study its movement.

Example 1.2 : *Script used to modify the pendulum's initial position*

```
//*****************************************************************
// animation of a spring pendulum
//*****************************************************************
//  function to create rotation matrix
function M=rot(a)
    M=[cos(a),sin(a);-sin(a),cos(a)];
endfunction
//  constants
n=40;      // number of coils of the spring
T=5;       // duration of the simulation
g=9.81;    //  g (gravitational acceleration)
k=10;      //  k (spring stiffness)
dt=0.01;   // dt (time step)

//*****************************************************************
// launch the graphics window
//*****************************************************************
//  window title
xtitle("(left-click to start the animation, right-click to stop)")
// title page (in LaTeX)
titlepage(["numerical solution of the spring pendulum ODE : ";" "; "$
$\Large r{d^2\over dt^2}a+2{d\over dt}r \times {d\over dt}a=g\times
 \sin(a)$$";
" "; "$$\Large {d^2\over dt^2}r-{k\over m}(r-r_0)=r\left({d\over dt}
 a\right)^2+g\times \cos(a)$$";" "; " with initial conditions : "; "$$
\Large  a(0)=? \;\;\;\;\;\; {d\over dt}a(0)=0  \;\;\;\;\;\; r(0)=r_0=?
 \;\;\;\;\;\; {d\over dt}r(0)=0 $$"])

//*****************************************************************
// processing the graphics window interactions
//*****************************************************************
// wait for a mouse click in the window
[c_i,c_x,c_y,c_w]=xclick();
while (c_i<>2)&(c_i<>5)   // as long as there is no right-click
      clf()    //clear the window
      //**********************************************************
      // retrieve the animation's initial data
      //**********************************************************
      // window title
      xtitle("(one click to initialize pendulum position a(0) and
 r(0) )")
      // parameterize the window's Axes handle
      plot(0,0,'.k');A=gca();A.x_location="origin";
```

```
A.y_location="origin";A.auto_scale="off";A.isoview="on";
A.data_bounds=[-1 -1; 1,0];xgrid(3)
// retrieve the pendulum's initial position coordinates :
[c_i,x,y,c_w]=xclick();
// compute initial values :
a=sign(x)*abs(atan(x/y));a0=a;da=0;  // compute angle a(0)
l=sqrt(x^2+y^2);r=l;,dr=0;           //  compute r(0)
// adapt the window's size to the pendulum's maximum size :
A.data_bounds=[-1.5,-max(4*l,4);1.5,max(l,0.5)];
//***********************************************************
// loop creates the animation
//***********************************************************
for t=0:dt:T
     //******************************************************
     // compute new positions
     //******************************************************
     // solve the differential equation for a and r using
     // Euler's method
     dda=-(g*sin(a)+2*dr*da)/r;
     ddr=r*(da)^2-k*(r-l)+g*cos(a);
     da=da+dt*dda;
     dr=dr+dt*ddr;
     a=a+dt*da;
     r=r+dt*dr;
     // compute the spring's line representation
     springr=linspace(0,r,n)';        // the spring stretches
     // coordinates transverse to spring's axis -> /\/\/\
     springa=[0;(-1).^[0:n-3]';0]*(l/10);
     //rotate the spring's picture by the angle a
     x=[x;r*sin(a)];
     y=[y;-r*cos(a)];
     M=-rot(-a);
     N=[springr,springa]*M;
     springy=N(:,1);springx=N(:,2);
     //******************************************************
     // display the pendulum
     //******************************************************
     drawlater()  // write to the graphics buffer
     clf()        // clear the window
     plot(springx,springy) //display the pendulum's spring
     xstring(0,0.1,["t=" string(t)]) // elapsed time
     // pendulum bob :
     xfarc(r*sin(a)-0.05,-r*cos(a)+0.05,0.1,0.1,0,360*64)
     // resize the graphics window
     A=gca();A.data_bounds=[-1.5,-max(4*l,4);1.5,max(l,0.5)];
     A.auto_scale="off";A.isoview="on";
     A.axes_visible=["off" "off" "off"];
     drawnow()             // display the graphics buffer
     realtime(t);           // delay display
end
//***********************************************************
// choose a new animation or exit program
//***********************************************************
xtitle("(one clic to proceed )")    // window title
plot(x,y,'-r')                       // display trajectory
A=gca();A.isoview="on";xgrid(3); // display grid (green)
```

```
    // waiting for a mouse click in graphics window :
    [c_i,x,y,c_w]=xclick();
    clf();                          // choose a new operation
    xtitle("(left-click to start the animations, right-click to
 stop)")
    plot(0,0,'.k');A=gca();A.x_location="origin";
    A.y_location="origin";
    // waiting for a mouse click in the window :
    [c_i,x,y,c_w]=xclick();
end
```

These two examples require knowledge of basic Scilab concepts that we are going to introduce in the rest of this book. Once you are done reading, you will be fully able to recreate them yourself. We will demonstrate this in the last chapter *Two Case Studies: a Pendulum and Comet Orbit*.

Caution › *To make sure the above scripts are properly executed, use Scilab version 5.4.1 or later.*

2

The Console

Launching Scilab opens up a window that consists of several elements. The first time it is opened after installation, the window should look like the one in Figure 2.1. The window's central section is the console. It allows the input of instructions next to the command prompt -->. These instructions are interpreted by Scilab following a carriage return (↵) and the result is displayed next to a new prompt.

Figure 2.1 : *Scilab's main window the first time it is launched*

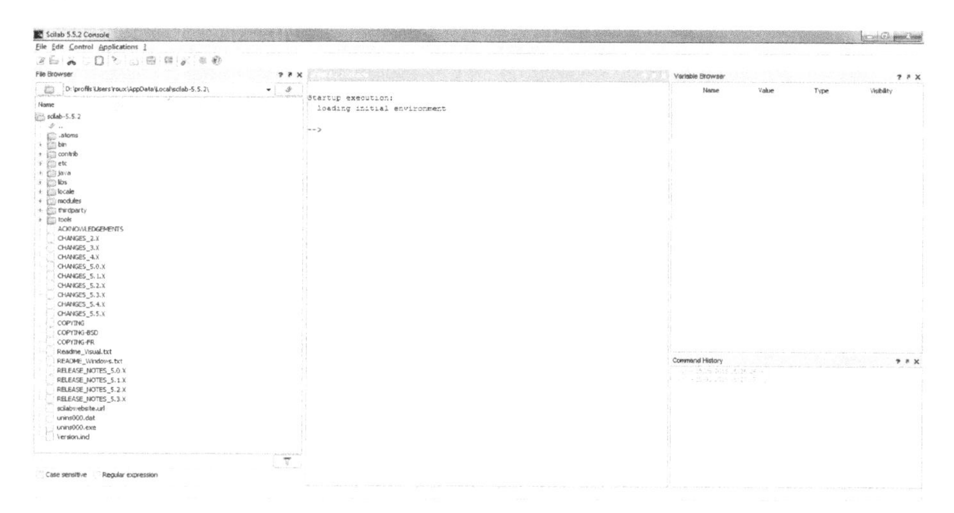

2.1. Taking ownership of the interface

The main window's subdivisions can be resized or made to appear in a window separate from the console. Other windows can also be added to the main window, as shown in the video in Figure 2.2. This convenient attribute of Scilab's graphical interface is called *docking*. This lets the user adjust the graphical interface to his/her own preferences and habits.

Figure 2.2 : Docking Scilab windows (video)

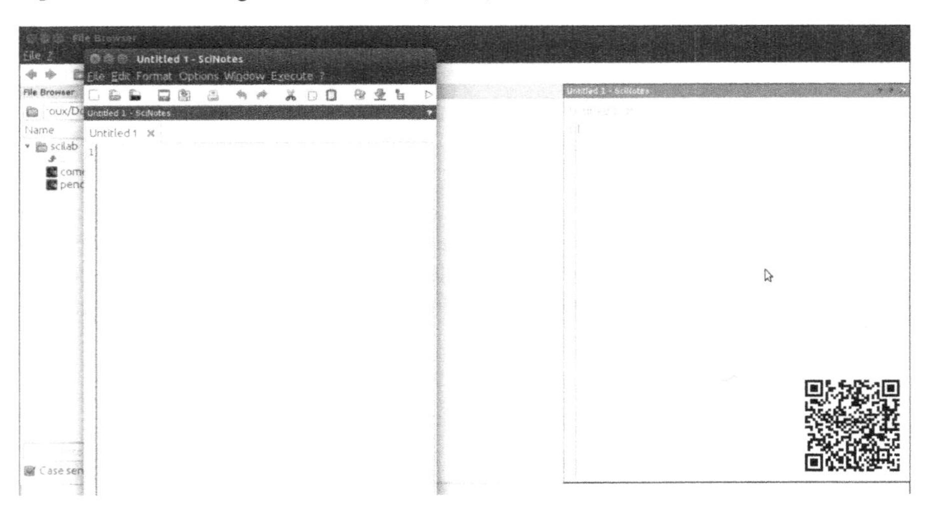

Tip › To dock a window to Scilab's main window, select its upper title bar with the mouse (once selected, the bar's color switches from gray to blue), then move it while maintaining the mouse right button pushed and release it where you wish to dock the window.

*Caution › When two windows are docked in the same location of the main window, they overlap and one can switch from one to the other by clicking on a tab. In the figure **below**, you can see that both the file browser and variable browser were docked to the same location and you can switch from one to the other by clicking on the tabs below the window.*

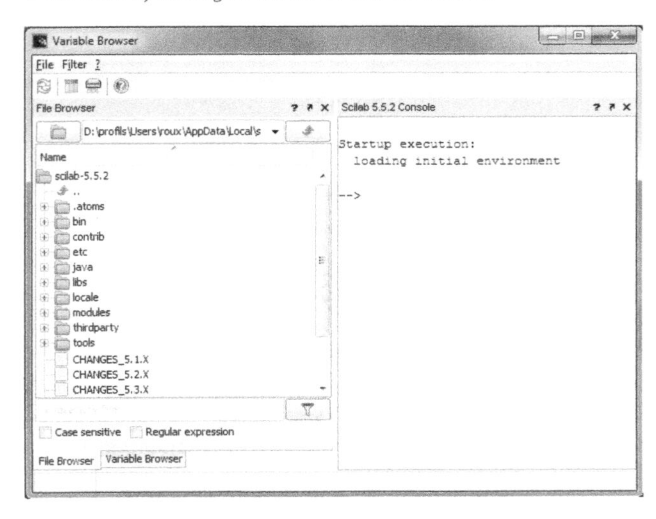

Since Scilab version 5.4.0, the preferred window configuration is saved at the end of each work session, which means that there is no need to rearrange the window each time Scilab is launched. You can also customize other interface settings by using the preference editor *via* the menu EDIT/PREFERENCES (See Figure 2.3). Among the different available parameters, you will find information on the localization (the language used for the Scilab interface).

Figure 2.3 : *Preference Editor*

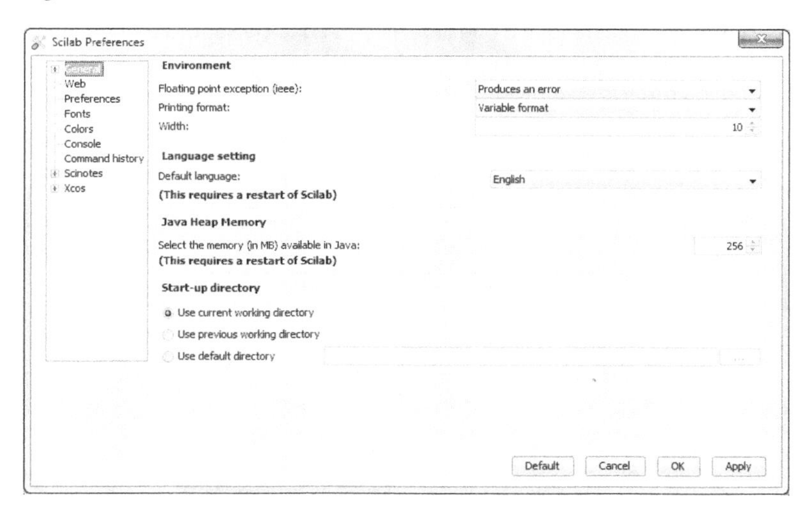

Tip › To explore the numerous possibilities Scilab offers, do not hesitate to browse the collection of demonstrations accessible from the ?/SCILAB DEMONSTRATIONS menu. A graphics window, shown in the figure **below**, *lets you explore the different files.*

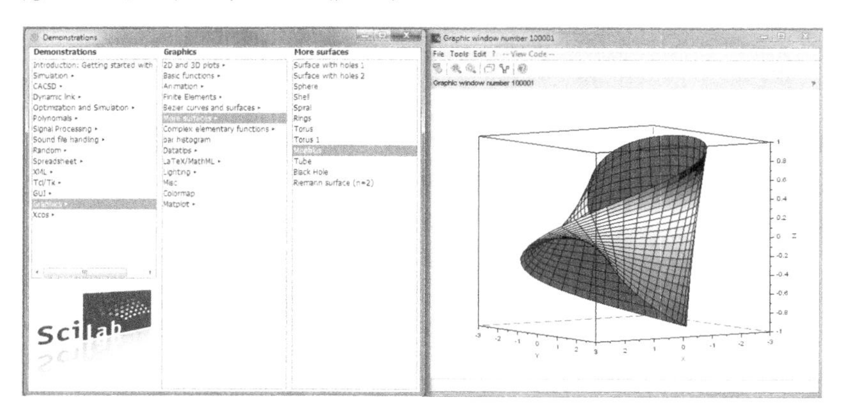

2.2. Using the console

The console is the most important Scilab window. It is a command prompt window similar to other command-line interpreters such as the Windows `cmd` or Linux `xterm`. It possesses the following list of functionalities typical of a command window.

- After the prompt `-->`, you can enter one or several commands. They are then processed after a carriage return, similar to the way a calculator operates. If you wish to write several commands on one line, they need to be separated by a comma (`,`) or a semicolon (`;`). Following a carriage return (↵), the commands are executed and the result is displayed.

```
-->2+2
 ans  =
    4.

-->2+2,3*4
 ans  =
    4.
 ans  =
    12.

-->2+2;3*4
 ans  =
    12.

-->2+2,3*4;      // a comment
 ans  =
    4.
```

*Tip › When the command you execute is followed by a semicolon, the result is not displayed in the console. The semicolon is very convenient to hide an intermediate calculation that would take up all the window space. Each time a result is displayed, the variable name in which it was stored is displayed before the equal sign (=). If no variable was specified, the result is by default stored in the variable **ans**, which stands for* answer *(also see Chapter **Variables, Constants and Types**).*

- You can also add comments on a command line by using a double slash (`//`): anything that follows will be ignored. Comments are especially useful when writing programs, as we will see in Part *Programming*.

- Navigating the command history can be achieved with the `uparrow` and `downarrow` of the keyboard. If one starts entering the first characters of a command, the command history navigation is limited to the lines beginning with those characters.

- The Tab key (⟶) enables the autocompletion of commands, as shown in Figure 2.4. When several commands start with the same characters, a pull-down menu lets you choose the desired command. You can also keep typing characters until autocomplete narrows down the options. Autocompletion also works to complete a variable name or a file/directory name.

Figure 2.4 : *Autocompletion of commands with Scilab*

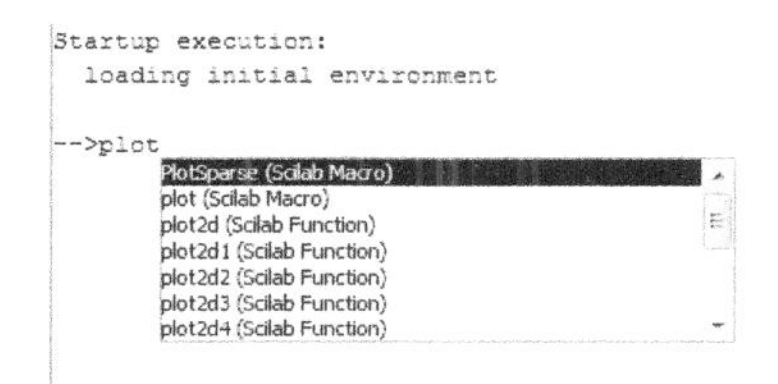

- It is also possible to copy/paste by right-clicking, as shown in Figure 2.5, or by using keyboard shortcuts (type `help console` to get a complete list of available keyboard shortcuts). Other functionalities can be accessed through the popup menu which appears when right-clicking.

Figure 2.5 : *Copy/paste with a right-click*

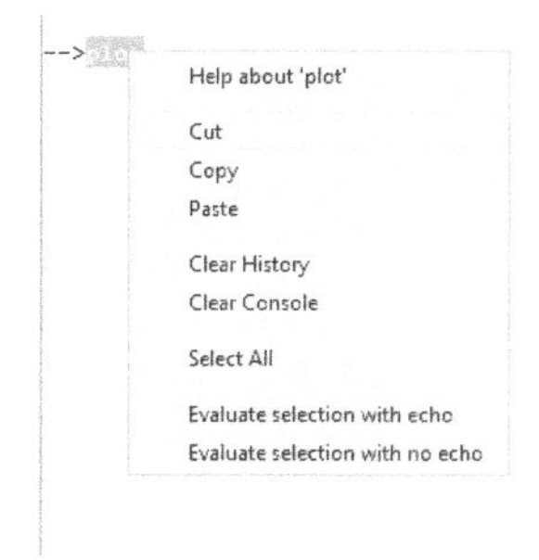

- The console can be cleared (without deleting the results of previously executed commands) by using the functions `clc` or `tohome`.

3

The Graphical Interface

In addition to the console, Scilab's graphical interface is made up of several windows which can be added to or removed from the main window (see Figure 2.2 video). In this chapter, you will learn the basic functionalities of Scilab's main windows.

3.1. The online help

The first window to be familiar with is, without doubt, the help browser window. It can be accessed in different ways:

- from the menu bar by clicking ? then SCILAB HELP
- by pressing the F1 key
- from the toolbar shortcut (see Figure 3.1)

Figure 3.1 : Toolbar shortcut for the online help

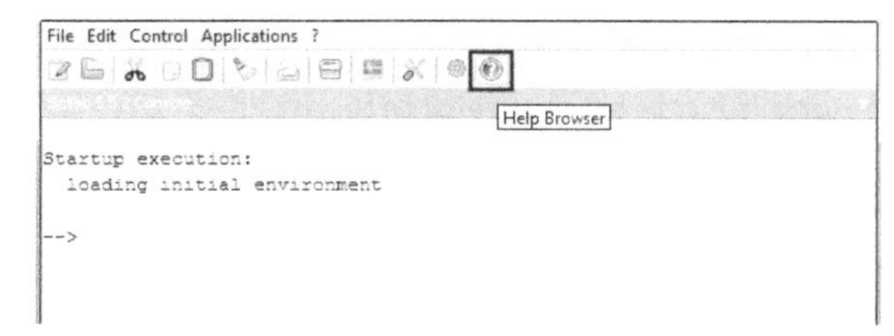

- from the console by entering one of the following commands:
  - `help` which lets you know what a command does and how to use it
  - `apropos` which lets you know which commands are relevant to a topic or keyword

```
-->apropos sinus
-->help sin
```

In either case, the window in Figure 3.2 opens up.

Figure 3.2 : Help window

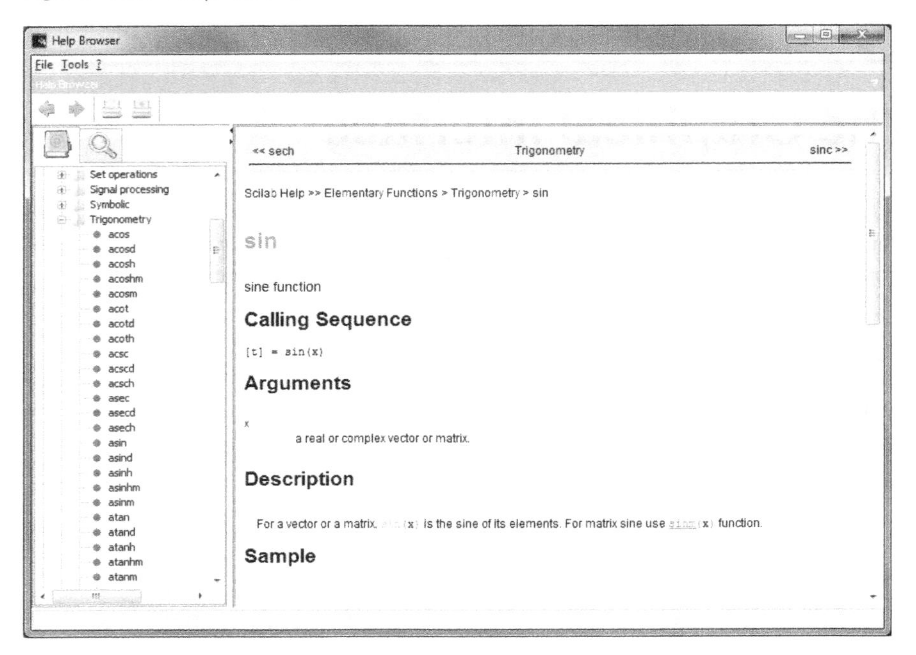

Tip › If you wish to get help on the **`sin`** *command, enter* **`help sin`**. *On the other hand, if you need help with the mathematical concept of* **`sine`**, *use* **`apropos sine`**. *In both cases, you should get a link redirecting you towards the* **`sine`** *function's help page.*

For each command, the help page is made up of several sections:

Calling sequence

Describes how to call a command to evaluate it in the Scilab console. In general, the calling sequence is of the form:

```
[output1,output2,...]=command(input1,input2,[option1,[option2]])
```

This sequence format means that `command` takes in two input values, indicated by `input1` and `input2`, and two optional values, `option1` et `option2`, enclosed in brackets `[` and `]`. The output values `output1` and `output2` can be retrieved by using brackets. There are several ways to call the command, for example:

- `ouput1=command(input1,input2)`
- `[output1,output2]=command(input1,input2)`

- `[output1,output2]=command(input1,input2,option1)`
- `output1=command(input1,input2,option1,option2)`
- `...`

*Caution › Whatever the syntax used, the value of **`output2`** can never be retrieved without the value of **`output1`**. Likewise, **`option2`** cannot be specified without first specifying **`option1`**.*

Arguments

Describes the type of arguments expected for the different values (`input1,input2,option1,... output2`) present in the calling sequence (see Chapter *Variables, Constants and Types*).

Description

Provides a more detailed explanation of what the `command` does.

Examples

Demonstrates how to use the `command` through examples of varying complexity. The source code displayed in the shaded box can be directly executed in the console or opened in the text editor by clicking the two buttons in the upper-right corner (See Figure 3.3).

Figure 3.3 : *Online help source code*

See also

Redirects to other help pages which may supplement the `command` help page.

3.2. The text editor

Scilab includes a text editor called SciNotes, which can be called in different ways:

- from the menu bar APPLICATIONS/SCINOTES
- from the Toolbar icon
- from the console with the command `editor` or `scinotes`

This editor possesses all the functionalities of a text editor intended for computer programming:

- semiautomatic formatting such as commenting/uncommenting parts of code, indentation of nested commands
- syntax coloring and command completion, autocompletion of closing parentheses
- search/replace function including the use of regular expressions
- several execution modes for the editor code (with/without echo) (see Figure 3.4)
- operation of Windows/Linux file formats (newlines CR and/or LF) and input and output character encodings (UTF-8, Latin1, ASCII, etc.)

Figure 3.4 : *SciNotes text editor*

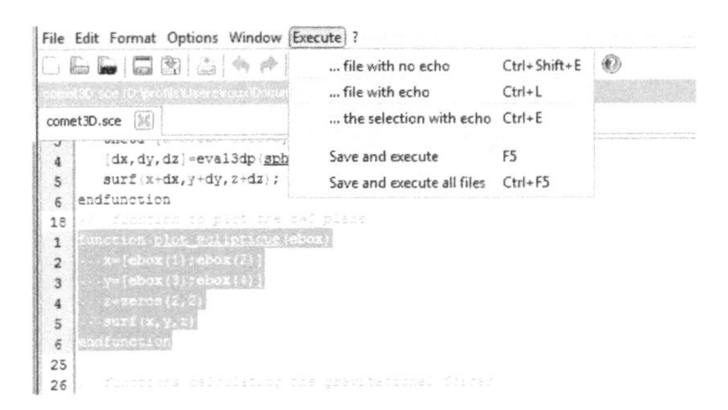

In the Part *Programming*, we will use this text editor to write programs in the Scilab language. You will discover various features used to control the launch and execution of Scilab programs in the editor and console graphical interfaces. They are accessible from the CONTROL menu. You can see an example of their use in the following video.

Figure 3.5 : Controlling the execution of Scilab programs (video)

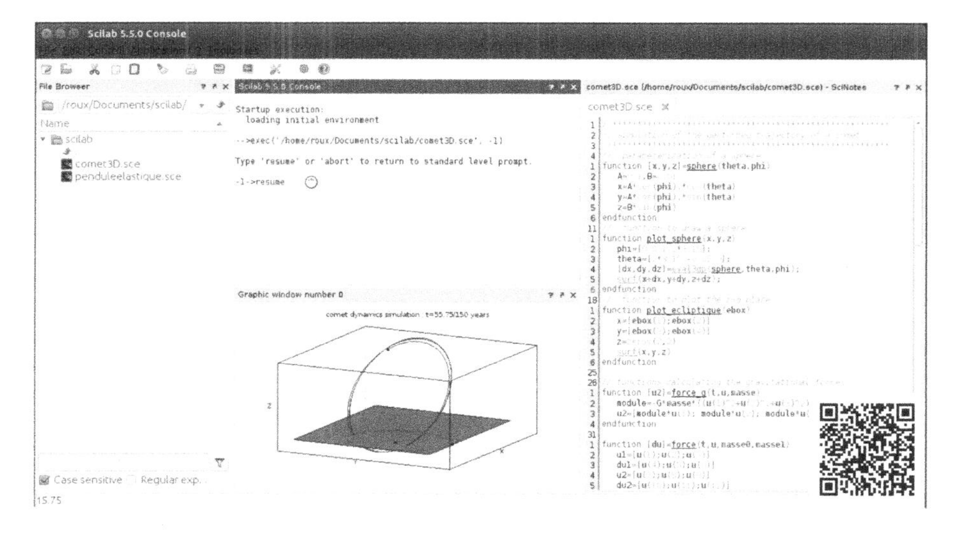

3.3. Other windows

Other Scilab windows have different specific purposes which are outlined here.

Graphics window

Scilab has a large range of graphical capabilities which, when used, cause a graphics window to open. For example, executing the `plot` command from the console opens up the graphics window shown in Figure 3.6.

Figure 3.6 : Scilab graphics window

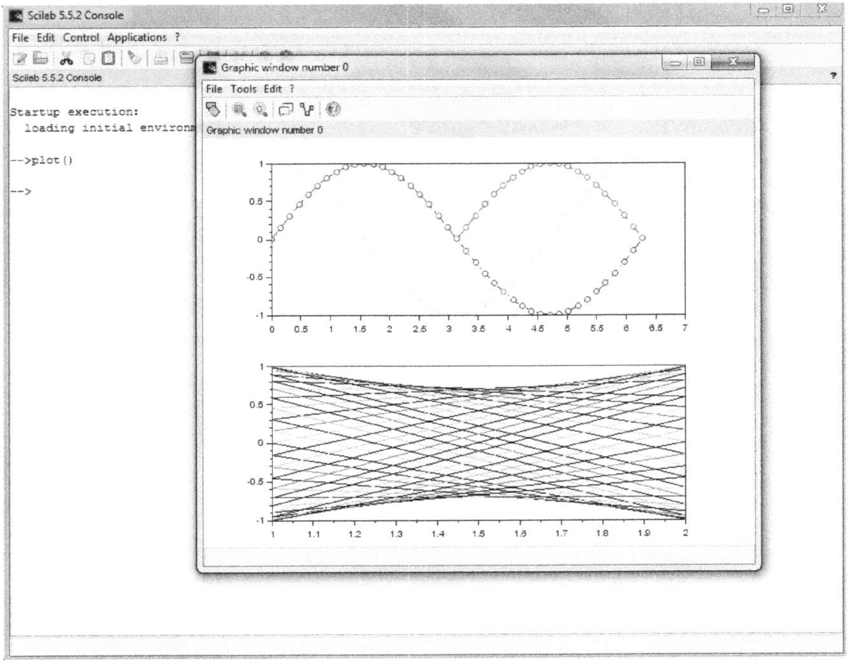

We will study Scilab's graphics properties in detail in Part *Creating Plots* and you will see how to create various figures in two as well as three dimensions! The graphics window lets the user interact easily with these figures. With the help of a mouse, you can for instance:

- zoom in and out, and restore the initial display scale
- modify the view angle for three-dimensional figures

The following video demonstrates how to use these capabilities.

Figure 3.7 : Zooming and modifying the view angle (video)

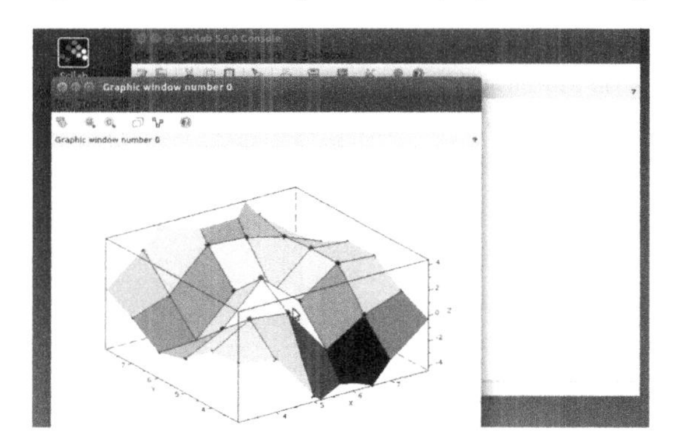

Scilab's graphics capabilities are based on a very detailed hierarchical display of graphics objects. This not only allows the creation of complex figures but also of new graphical interfaces (see Section 23.4, *Creating your own graphical interfaces*). A graphics editor, accessible from the menu bar in the graphics window (menu EDIT then FIGURE PROPERTIES), lets you navigate and modify the different parameters that constitute the graph. The graphics window incorporates modifications in real time (see the following video).

Figure 3.8 : Using the graphics editor (video)

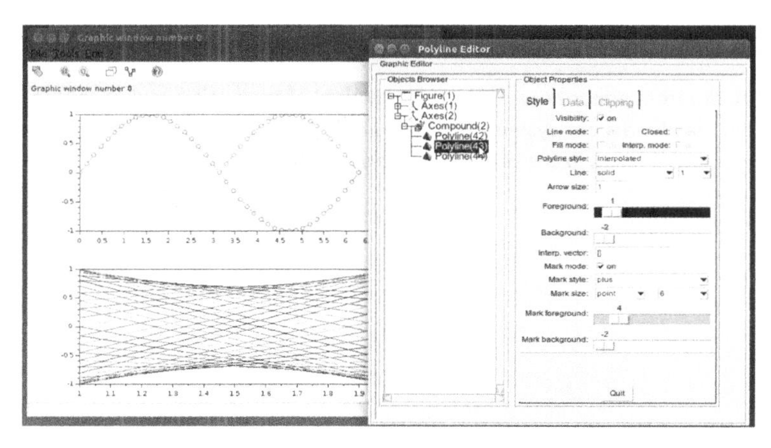

Caution › The graphics editor currently does not work on the Scilab Mac OS versions.

ATOMS

The external module manager ATOMS lets you install or uninstall supplementary modules related to Scilab. These modules add advanced capabilities to the software, for example SIVP can be used for image processing or Metanet for graphing and network computations (see Section 5.4, *Supplementary modules on Forge*).

Figure 3.9 : *Choosing the SIVP supplementary module from the ATOMS window*

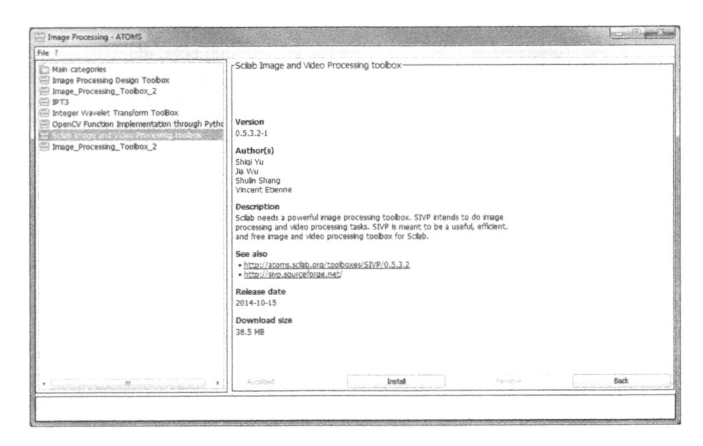

Xcos

The bloc-diagram editor Xcos, Scilab's equivalent of Matlab's Simulink toolbox, lets you simulate dynamical systems. For more information on Xcos, see the manuel [http://d-booker.jo.my/xcos-book] that is dedicated to it.

Figure 3.10 : *Xcos diagram of a vehicle suspension*

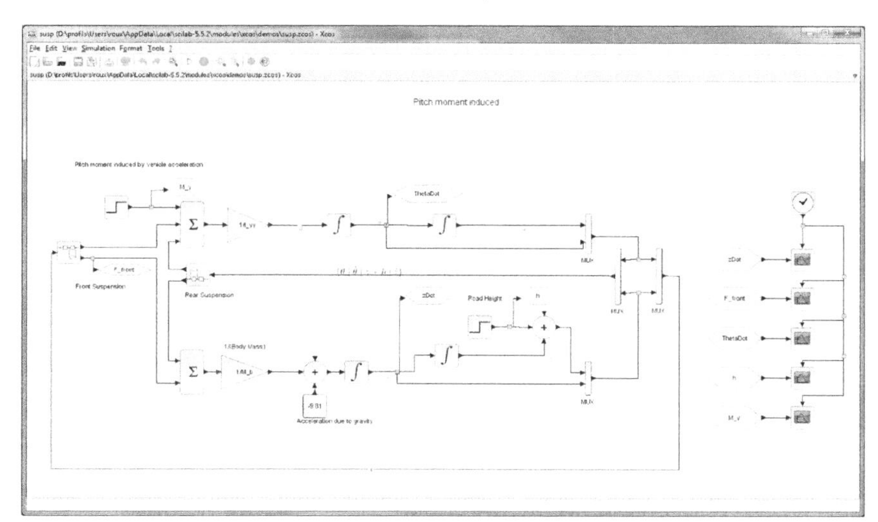

Various tool managers

Several managers have been added to Scilab since version 5: a history manager, a variable editor and a file browser. They are loaded by default the first time Scilab is launched (see Figure 2.1). You can also launch them from the console with the commands: `filebrowser`, `browsevar`, `editvar` and `browsehistory`.

4

Inputs/Outputs

In this chapter, you will see how Scilab integrates itself inside the computer environment. You will be given examples of simple commands executed in the console. These will help you get accustomed to using the console.

4.1. File system

The first concept that needs to be understood to use Scilab with ease is that of the current directory. At each moment, a directory of your file system gets associated to your work environment in Scilab. By default, this is the directory Scilab will search in to find auxiliary information during certain operations (e.g. opening files, writing to files). There exist several ways to interact with the file system:

- retrieve the name of the current directory through the FILE menu in the console or with the command `pwd`
- change the current directory through the FILE menu in the console or by using the commands `cd` or `chdir`
- create a new directory with `mkdir` and remove a directory with `rmdir`
- move, copy or delete directories straight from the console with the commands `copyfile`, `movefile` and `mdelete`

Paths for certain Scilab-specific directories are accessible from the console with the following commands:

- Scilab's installation directory can be retrieved with `SCI` (also see Section 6.2, *Installation*).
- The directory that stores user data (history, preferences, etc) is displayed with `SCIHOME`. This directory will change depending on the operating system used and the way the user accounts are managed. For example, the directory will be, in general:
  - `C:/Users/<User>/AppData/Roaming/Scilab/<Scilab-Version>` for Windows

   - `/home/<User>/.Scilab/<Scilab-Version>` for Unix-type systems
   - `/Users/<User>/.Scilab/<Scilab-Version>` for Mac OS

- A temporary directory is also assigned to each Scilab work session. It is created at the beginning of the session and destroyed at the end. Its path can be retrieved from the console with `TMPDIR`. Its name is of the form `SCI_TMP_*` and its location depends on the operating system.

Here are a few examples of directory operations that you can perform from the console:

```
-->path=pwd();  // current directory

-->cd SCI  // go to the Scilab installation directory
 ans  =
 D:\profils\Users\roux\AppData\Local\scilab-5.5.2

-->pwd  // value of current directory
 ans  =
 D:\profils\Users\roux\AppData\Local\scilab-5.5.2

-->cd contrib  // go to the SCI/contrib/ directory
 ans  =
 D:\profils\Users\roux\AppData\Local\scilab-5.5.2\contrib

-->cd '../'  // "move up" to the SCI directory
 ans  =
 D:\profils\Users\roux\AppData\Local\scilab-5.5.2

-->chdir('contrib')  // go to the SCI/contrib/ directory
 ans  =
  T

-->pwd  // value of current directory
 ans  =
 D:\profils\Users\roux\AppData\Local\scilab-5.5.2\contrib

-->chdir(TMPDIR)  // go to the temporary directory
 ans  =
  T

-->mkdir('test')  // create the directory test/
 ans  =
    1.

-->ls('te*') // list the elements starting with "te"
 ans  =
 test

-->rmdir('test')  // remove the test/ directory
 ans  =
    1.
```

```
-->dir('te*') // this is the content of the current directory which is
 empty []
 ans  =
[]

-->chdir(path)  // return to the initial current directory
 ans  =
  T
```

All these operations can also be performed *via* a graphics interface through the file browser (see Figure 4.1). This browser can be called from the APPLICATIONS menu from the console or with the command `filebrowser`.

Figure 4.1 : *File browser*

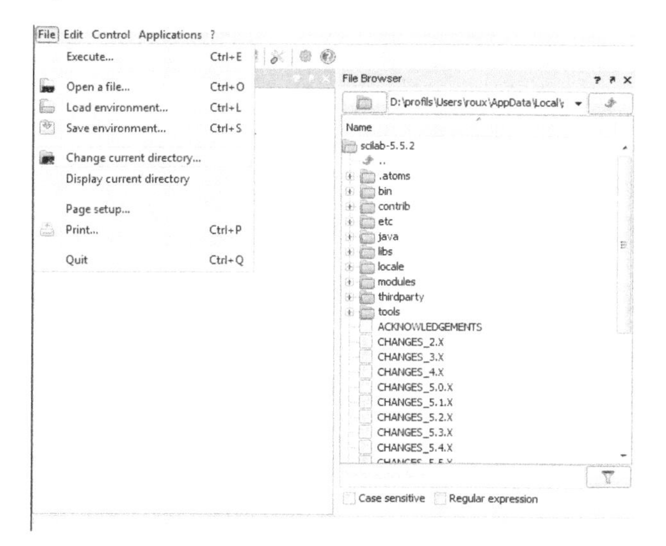

*Tip › Scilab's launch shortcut can be customized to specify the startup directory. By default, this directory in usually the root directory of the user launching Scilab or the Scilab installation directory (**SCI**) for Windows.*

4.2. System commands

It is possible to call system commands from Scilab. Depending on your operating system you may use the command `dos` or `unix`, but both commands work the same way! Four versions of the command `unix` process the result returned by the system in different ways:

- `unix_g` lets you redirect the output to a Scilab variable.

- `unix_w` redirects the output to the console.
- `unix_x` redirects the output to a popup window (see Figure 4.2).
- `unix_s` does not return anything.

Here are several examples of results displayed in the console:

```
-->unix('dir')       // return code
 ans  =

    0.

-->unix_s('dir')     // no output

-->unix_g('dir')     // output to variable
 ans  =

! Volume in drive C has no label.                                    !
!                                                                    !
! Volume Serial Number is 3825-15B6                                  !
!                                                                    !
!                                                                    !
!                                                                    !
! Directory of C:\Program Files\scilab-5.5.0\contrib                 !
!                                                                    !
!                                                                    !
!                                                                    !
!09/14/2014  02:05 PM    <DIR>          .                            !
!                                                                    !
!09/14/2014  02:05 PM    <DIR>          ..                           !
!                                                                    !
!04/11/2014  02:03 AM               119 loader.sce                   !
!                                                                    !
!09/14/2014  02:05 PM    <DIR>          toolbox_skeleton             !
!                                                                    !
!09/14/2014  02:05 PM    <DIR>          xcos_toolbox_skeleton        !
!                                                                    !
!               1 File(s)            119 bytes                       !
!                                                                    !
!               4 Dir(s)  364,507,512,832 bytes free                 !

-->unix_w('dir')     // output to console
 Volume in drive C has no label.
 Volume Serial Number is 3825-15B6

 Directory of C:\Program Files\scilab-5.5.0\contrib

09/14/2014  02:05 PM    <DIR>          .
09/14/2014  02:05 PM    <DIR>          ..
04/11/2014  02:03 AM               119 loader.sce
09/14/2014  02:05 PM    <DIR>          toolbox_skeleton
09/14/2014  02:05 PM    <DIR>          xcos_toolbox_skeleton
               1 File(s)            119 bytes
               4 Dir(s)  364,506,988,544 bytes free
```

```
-->unix_x('dir')     // output to window

-->cd(path);
```

Caution › *The* `unix` *command returns an integer code which depends on the given result.*

Figure 4.2 : *Example of popup window returned by* `unix_x`

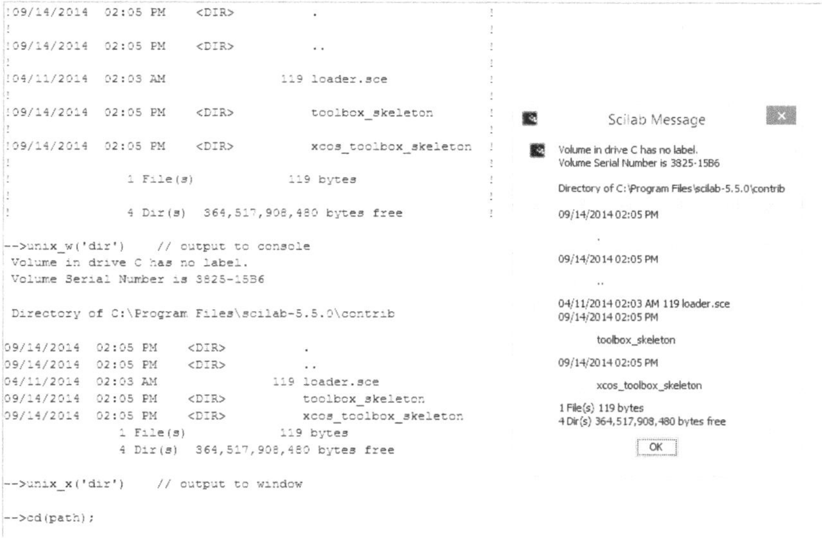

```
!09/14/2014  02:05 PM    <DIR>          .                        !
!                                                                !
!09/14/2014  02:05 PM    <DIR>          ..                       !
!                                                                !
!04/11/2014  02:03 AM              119 loader.sce                !
!                                                                !
!09/14/2014  02:05 PM    <DIR>          toolbox_skeleton         !
!                                                                !
!09/14/2014  02:05 PM    <DIR>          xcos_toolbox_skeleton    !
!                                                                !
!               1 File(s)            119 bytes                   !
!                                                                !
!               4 Dir(s)  364,517,908,480 bytes free             !

-->unix_w('dir')     // output to console
 Volume in drive C has no label.
 Volume Serial Number is 3825-15B6

 Directory of C:\Program Files\scilab-5.5.0\contrib

09/14/2014  02:05 PM    <DIR>          .
09/14/2014  02:05 PM    <DIR>          ..
04/11/2014  02:03 AM              119 loader.sce
09/14/2014  02:05 PM    <DIR>          toolbox_skeleton
09/14/2014  02:05 PM    <DIR>          xcos_toolbox_skeleton
               1 File(s)            119 bytes
               4 Dir(s)  364,517,908,480 bytes free

-->unix_x('dir')     // output to window

-->cd(path);
```

```
Scilab Message
Volume in drive C has no label.
Volume Serial Number is 3825-15B6

Directory of C:\Program Files\scilab-5.5.0\contrib

09/14/2014 02:05 PM
.
09/14/2014 02:05 PM
..
04/11/2014 02:03 AM 119 loader.sce
09/14/2014 02:05 PM
toolbox_skeleton
09/14/2014 02:05 PM
xcos_toolbox_skeleton
1 File(s) 119 bytes
4 Dir(s) 364,517,908,480 bytes free
OK
```

To retrieve information related to the operating system that's running Scilab, use the command `getos`. Similarly, the command `getversion` tells you which Scilab version is being used. More broadly, an operating system variable can be retrieved with the command `getenv`.

```
-->getversion()        // Scilab version
 ans  =
 scilab-5.5.2

-->getos()            // os windows
 ans  =
 Windows

-->getenv('TMP')       // retrieve the environment variable TMP
 ans  =
 D:\profils\Users\roux\AppData\Local\Temp
```

Moreover, you can interact with the clipboard by using `clipboard`.

```
-->clipboard("copy","test")     // CTRL+C the text "test"
 ans  =
    []

-->clipboard("paste")           // CTRL+V
 ans  =
 test
```

4.3. CPU dates and times

For certain applications, you may need to access information related to time. Different commands can be used to evaluate durations. You can:

- calculate the CPU time (= number of processor cycles) elapsed between two operations by using `timer`
- calculate the real elapsed time in milliseconds between two actions with the command `tic`/`toc`, which starts/stops the Scilab timer
- pause Scilab for a certain duration with `sleep(time in milliseconds)` or `xpause(time in microseconds)`
- perform real-time simulations with `realtimeinit` (which allows you to set the time unit) and `realtime`. The first call to `realtime` sets the current date origin and subsequent calls force Scilab to wait until a specified date has passed to carry on.

Test these different commands with the following examples:

```
-->// time with tic() and toc()

-->tic()

-->sleep(1000)    // 1000ms=1second

-->toc()
 ans  =
    1.005


-->tic()

-->xpause(200000)    // 200000micros=0.2 seconds

-->toc()
 ans  =
    0.206
```

```
-->// CPU time with timer

-->timer();

-->sleep(1000)    // 1000ms=1second

-->timer()
 ans  =
    0.0468003

-->timer();

-->xpause(1000000)    // 1000000micros=1 second

-->timer()
 ans  =
    0.0312002

-->//real time

-->realtimeinit(1)    // time unit of 1 second

-->realtime(0)    // sets current date to t=0

-->tic()

-->realtime(2)    // wait for date t=2

-->toc()    // the timer will display 2 secondes
 ans  =
    2.003
```

You can perform calculations using dates, but be careful, numerous formats can be used to manage dates and as many functions exist to manipulate them.

- `clock` retrieves the date as a vector with six parameters `[year, month, day, hour, minute, second]`.
- `datenum` retrieves a date in the form of the number of days elapsed since January 1st of year zero.
- `getdate` retrieves a date as a timestamp (number of seconds elapsed since January 1st 1970) or as a vector of ten parameters `[year, month, week, Julian day, day of the week, day of the month, hour, minute, second, millisecond]` (more complex than the `clock` outputs!).

Once the dates are retrieved, they can be processed by using other commands:

- `datevec` converts a timestamp into a vector corresponding to the appropriate date for Scilab.
- `weekday` computes the day of the week corresponding to a timestamp.

- `eomday` computes the last day of a certain month for a given year.
- `etime` calculates the difference (in seconds) between two dates given by a vector of six parameters.

Finally, the following functions display dates or calendars:

- `date` retrieves a date as a character string.
- `calendar` displays a monthly or annual calendar.

Caution › For the function **weekday**, *the first day of the week is Sunday, whereas for the function* **calendar**, *the first day of the week is Monday. If no arguments are specified, these functions return the values corresponding to the current date. Otherwise, they return results for the dates specified as parameters.*

Here are a few examples of use of the previous functions:

```
-->calendar(1970,1)     // Jan. 1970 calendar
 ans  =

       ans(1)

 Jan 1970

       ans(2)

    M       Tu      W       Th      F      Sat      Sun

       ans(3)

    0.      0.      0.      1.      2.      3.      4.
    5.      6.      7.      8.      9.      10.     11.
    12.     13.     14.     15.     16.     17.     18.
    19.     20.     21.     22.     23.     24.     25.
    26.     27.     28.     29.     30.     31.     0.
    0.      0.      0.      0.      0.      0.      0.

-->eomday(2012,2)     // last day of February 2012
 ans  =
    29.

-->d1=[1970 1 1 0 0 0]     // Scilab date format
 d1  =
    1970.    1.    1.    0.    0.    0.

-->t1=datenum(d1)     // serial date number for date d1
 t1  =
    719529.

-->[N,S]=weekday(t1)     // day of the week for date d1
 S  =
 Thu
```

```
  N  =
    5.

-->date()     // current date
 ans  =
 16-Jun-2015

-->d2=clock()     // scilab vector for current date
 d2  =
    2015.    6.    16.    21.    11.    18.000004

-->t2=datenum(d2)     // serial date number of date d2
 t2  =
    736131.88

-->datevec(t2)     // date corresponding to t2
 ans  =
    2015.    6.    16.    21.    11.    18.000004

-->etime(d1,d2)     // difference between dates d1 and d2
 ans  =
  - 1.434D+09

-->d=etime(d2,d1)     // difference between dates d2 and d1
 d  =
    1.434D+09

-->getdate(d)     // day occurring d seconds after Jan. 1st 1970
 ans  =
    2015.    6.    25.    167.    3.    16.    23.    11.    18.
 0.0000038
```

4.4. Command history

Scilab provides a command history browser which automatically records commands sent to the console. This capability helps speed up the script writing process. With the help of the following functions, you will be able to:

- clear the command history with `resethistory`
- save it to a file with `savehistory`, reload it in the Scilab environment with `loadhistory`
- modify entries in the history with `addhistory` and `removelinehistory`
- retrieve and store it in a variable with `gethistory`
- display the command history within the browser with `browserhistory` or within the console with `displayhistory`

You can also use the history browser (see Figure 4.3) if you prefer to use a graphical interface to perform the operations described above.

Figure 4.3 : *The history browser*

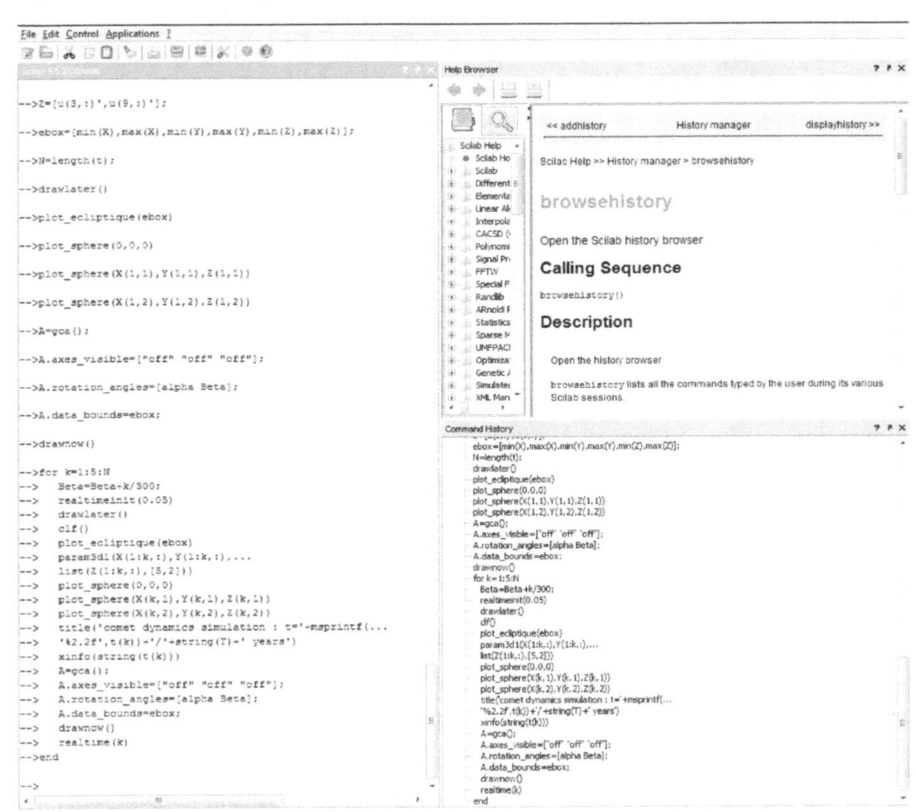

Finally, the function `diary` lets the user store anything displayed in the console inside a text file. Commands, along with their results, if applicable, are recorded between two calls to the function `diary`.

```
-->path=pwd();                    // current directory

-->id=diary('scilab-base-diary.txt')    // open the diary
 id  =
    2.

-->cd TMPDIR                      // temporary Scilab directory
 ans  =
 D:\profils\Users\roux\AppData\Local\Temp\SCI_TMP_6184_

-->resethistory()                 // erase history log
```

```
-->addhistory('ls')            // add a line to the history log

-->gethistory()                // retrieve the command history inside a
 variable
 ans  =
!// -- 16/06/2015 21:11:16 -- //  !
!                                 !
!ls                               !

-->displayhistory()            // display history log
0 : // -- 16/06/2015 21:11:16 -- //
1 : ls

-->savehistory('essai.txt')  // save the history inside a file

-->dir('essai.txt')            // the file is created inside the current
 directory
 ans  =
essai.txt

-->browsehistory()             // open the history browser

-->cd(path);                   // return to the initial directory

-->diary(id,'close')           // close the diary
```

Caution* › *When a Scilab program is launched from a saved file (like the ones shown in the chapter* **Preview of Scilab) by using graphical interface shortcuts (see the video in* **Figure** *3.5), the commands executed, including the* **exec** *command that appears in the console, are left out of the history log.***

5

Finding Information on Scilab

Learning a new language or software by yourself usually requires outside help. The Internet can provide you with that help in different ways. In this chapter, you will find several useful sites you can visit to find information on Scilab.

5.1. Documentation on the Scilab website

The Scilab website is your starting point. It contains a Documentation [http://d-booker.jo.my/sci-documentation-en] section in which you will find:

- all the Scilab help pages that describe each Scilab feature (these same pages are accessible from the software)
- a wiki with information on certain developmental procedures specific to Scilab (developing supplementary modules, using the Scilab API, porting MATLAB tools to Scilab)
- a MATLAB-Scilab dictionary
- tutorials, books and articles, etc.

The official Scilab website also houses a page [http://d-booker.jo.my/sci-fileexchange] dedicated to the user exchange of Scilab programs.

Figure 5.1 : Scilab File Exchange page

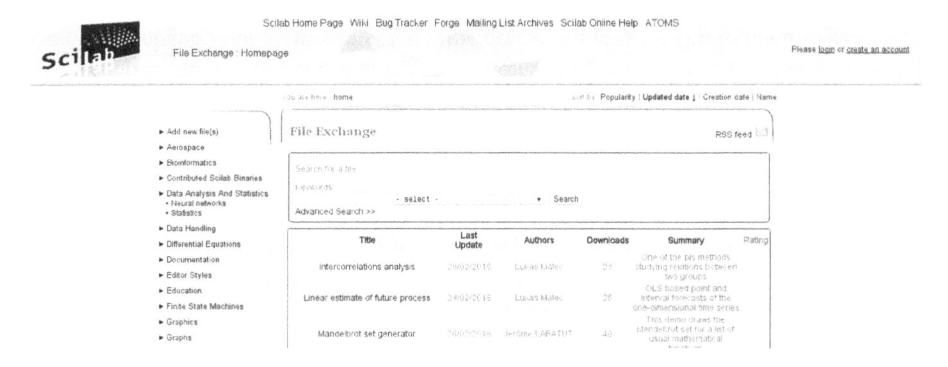

5.2. Mailing Lists

For those who wish to get more personalized assistance, there exist numerous mailing lists dedicated to Scilab. They facilitate exchanges between Scilab users and developers. These official lists [http://d-booker.jo.my/sci-ml-en] are located on the scilab.org website. You can find lists specifically made for users:

- a list for english-speakers users@lists.scilab.org [http://d-booker.jo.my/sci-ml-users]
- a list for french-speakers users-fr@lists.scilab.org [http://d-booker.jo.my/sci-ml-users-fr]
- for questions pertaining to the educational use of Scilab, see enseignement@lists.scilab.org [http://d-booker.jo.my/sci-ml-ens] (exchanges in French)

and lists (only in English) dedicated to developers:

- the primary list dev@lists.scilab.org [http://d-booker.jo.my/sci-ml-dev]
- for issues related to the translation of the Scilab interface in different languages localization@lists.scilab.org [http://d-booker.jo.my/sci-ml-localization]
- for questions linked to the inclusion of Scilab in a distribution distributor@lists.scilab.org [http://d-booker.jo.my/sc-ml-distribution]

There are also other mailing lists, such as usenet [http://d-booker.jo.my/sci-usenet], or forums, however these websites are not managed by the Scilab development team.

5.3. Keeping track of bugs with Bugzilla

The development and growth of the Scilab software is based on the continuous dialog between users and developers. To ensure the proper management of user requests, the Scilab team uses a bug tracking system called Bugzilla, which is accessible at http://bugzilla.scilab.org/.

When you discover an issue while using Scilab, it is generally advisable to report the incident on Bugzilla, unless the issue has already been identified. This helps the development team improve the software and increases the chances of getting the issue you found fixed. In order to do this, you need to create a user account on Bugzilla. Then, you need to fill out a form detailing the issue and, if possible, include an example of a way to reproduce the problem.

Figure 5.2 : Scilab's Bug Tracker

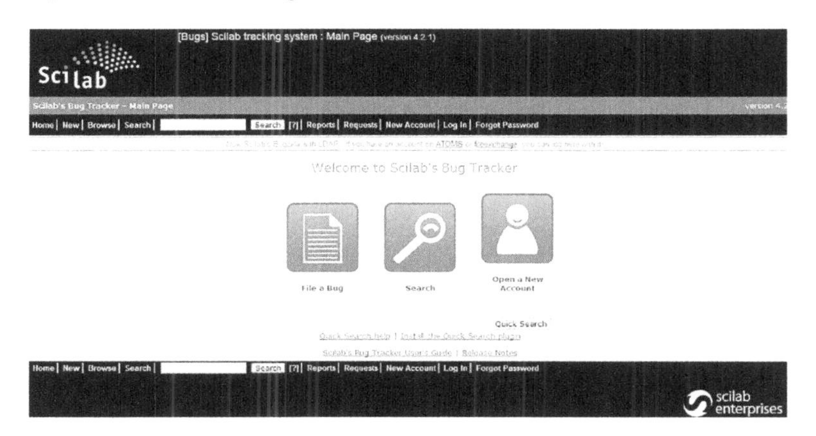

Figure 5.3 : Reporting a bug on Bugzilla

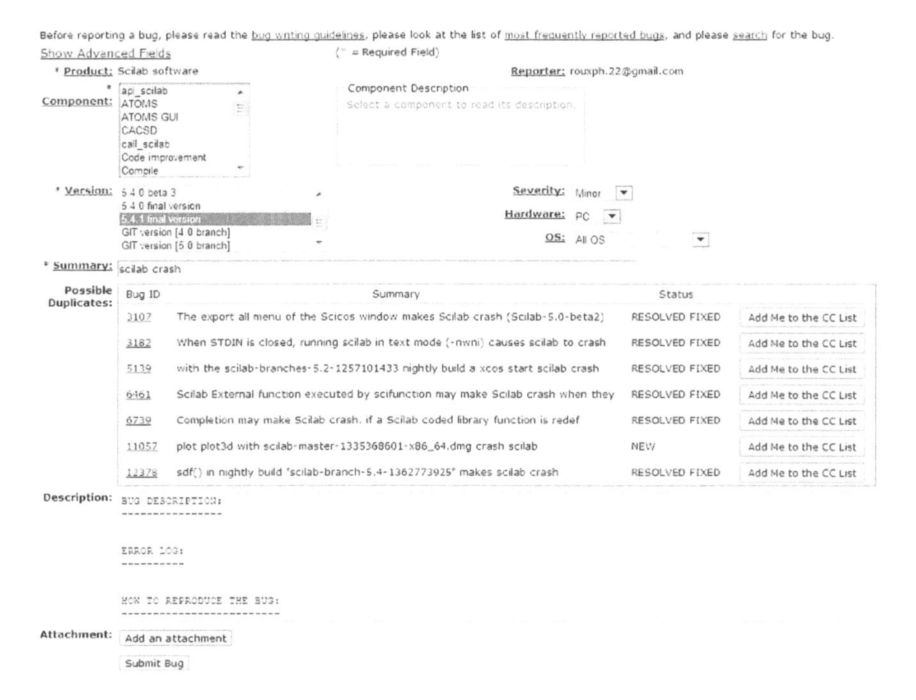

***Tip** › Certain bugs may be linked to your operating system or the libraries used by your machine. In order to determine what the bug is linked to, provide information on the environment in which the bug was found. The command **ver** lets you easily retrieve this information in the Scilab console:*

```
-->ver()
 ans  =
!Scilab Version:                     5.5.2.1427793548                  !
!                                                                       !
!Operating System:                   Windows 7 6.1                     !
!                                                                       !
!Java version:                       1.6.0_41                          !
!                                                                       !
!Java runtime information:           Java(TM) SE Runtime Environment
                                     (build 1.6.0_41-b02)              !
!                                                                       !
!Java Virtual Machine information:   Java HotSpot(TM) 64-Bit Server
                                     VM (build 20.14-b01, mixed mode)!
!                                                                       !
!Vendor specification:               Sun Microsystems Inc.            !
```

***Caution** › Before reporting a bug, make sure the bug is not already listed by searching the Bugzilla reports list. You can search the database straight from Bugzilla's main page by using keywords, as shown in the figure **below** (with the keyword surf).*

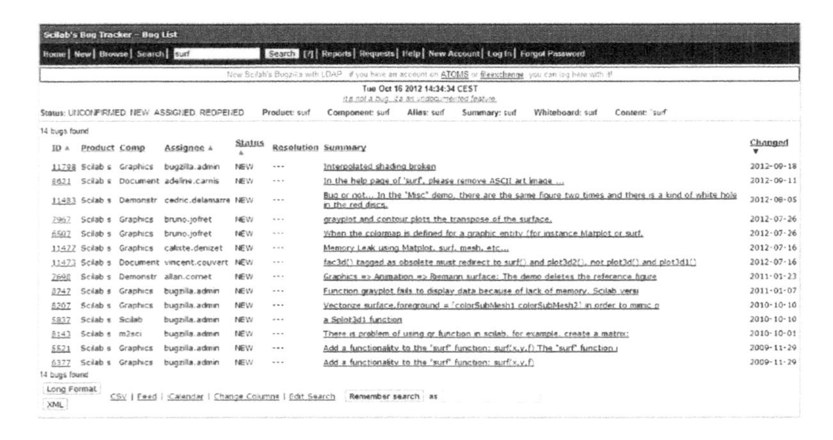

Scilab's Bug Tracker – Bug List

Home | New | Browse | Search | surf Search [?] | Reports | Requests | Help | New Account | Log In | Forgot Password

New Scilab's Bugzilla with LDAP: if you have an account on ATOMS or fileexchange, you can log here with it!

Tue Oct 16 2012 14:34:34 CEST

It's not a bug, it's an undocumented feature.

Status: UNCONFIRMED NEW ASSIGNED REOPENED Product: surf Component: surf Alias: surf Summary: surf Whiteboard: surf Content: "surf"

14 bugs found

ID	Product	Comp	Assignee	Status	Resolution	Summary	Changed
11798	Scilab s	Graphics	bugzilla.admin	NEW	---	Interpolated shading broken	2012-09-18
8621	Scilab s	Document	adeline.carnis	NEW	---	In the help page of 'surf', please remove ASCII art image ...	2012-09-11
11483	Scilab s	Demonstr	cedric.delamarre	NEW	---	Bug or not... In the "Misc" demo, there are the same figure two times and there is a kind of white hole in the red discs.	2012-08-05
7967	Scilab s	Graphics	bruno.jofret	NEW	---	grayplot and contour plots the transpose of the surface.	2012-07-26
6507	Scilab s	Graphics	bruno.jofret	NEW	---	When the colormap is defined for a graphic entity (for instance Matplot or surf,	2012-07-26
11477	Scilab s	Graphics	calixte.denizet	NEW	---	Memory Leak using Matplot, surf, mesh, etc...	2012-07-16
11473	Scilab s	Document	vincent.couvert	NEW	---	fac3d() tagged as obsolete must redirect to surf() and plot3d2(), not plot3d() and plot3d1()	2012-07-16
7698	Scilab s	Demonstr	allan.cornet	NEW	---	Graphics => Animation => Riemann surface: The demo deletes the referance figure	2011-01-23
8747	Scilab s	Graphics	bugzilla.admin	NEW	---	Function grayplot fails to display data because of lack of memory. Scilab versi	2011-01-07
8207	Scilab s	Graphics	bugzilla.admin	NEW	---	Vectorize surface.foreground = ['colorSubMesh1 colorSubMesh2'] in order to mimic p	2010-10-10
5837	Scilab s	Scilab	bugzilla.admin	NEW	---	a Splot3d1 function	2010-10-10
8143	Scilab s	m2sci	bugzilla.admin	NEW	---	There is problem of using qr function in scilab. for example, create a matrix:	2010-10-01
5521	Scilab s	Graphics	bugzilla.admin	NEW	---	Add a functionality to the 'surf' function: surf(x,y,f) The "surf" function i	2009-11-29
6377	Scilab s	Graphics	bugzilla.admin	NEW	---	Add a functionality to the 'surf' function: surf(x,y,f)	2009-11-29

14 bugs found

Long Format XML CSV | Feed | iCalendar | Change Columns | Edit Search Remember search as

5.4. Supplementary modules on Forge

As we previously mentioned in the chapter *The Graphical Interface*, Scilab's capabilities can be enhanced by adding supplementary modules developed for specific applications. For example:

- Metanet to manage graphs and networks

- SIVP for image processing
- Guimaker to create graphical interfaces
- Scimax for symbolic computation (*via* the software Maxima)

These modules are individual projects, however they are completely dependent on Scilab. There exist a lot of projects such as these (more than a hundred) that can all be found on the Scilab Forge [http://d-booker.jo.my/sci-forge] (see Figure 5.4), which facilitates searches.

Figure 5.4 : *Scilab Forge page for the Metanet supplementary module*

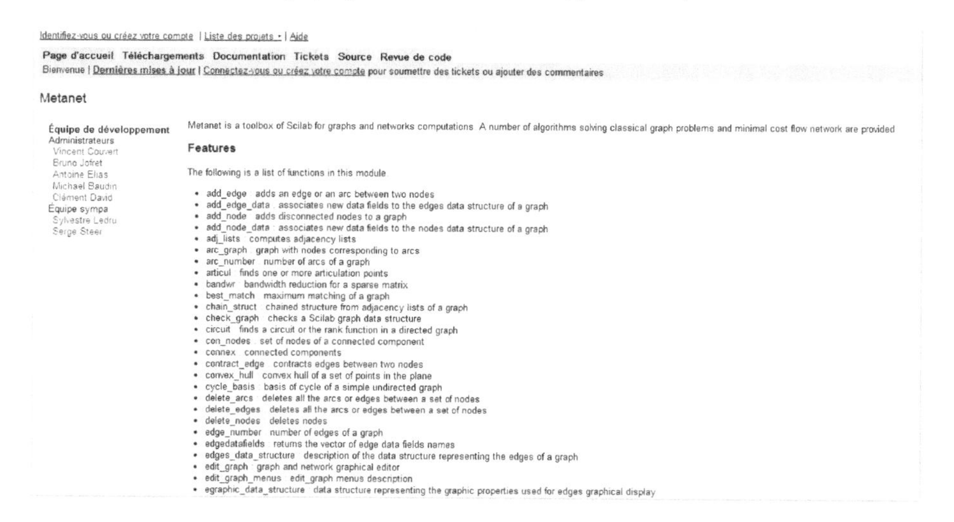

Each supplementary module's page in Forge lets users access source codes as well as report bugs. You can install/uninstall these modules *via* the module manager ATOMS or by using the commands:

- `atomsInstall` to install a module (see Figure 5.5)
- `atomsRemove` to uninstall a module

Figure 5.5 : Installing a supplementary module with `atomsInstall`

```
File Edit Control Applications ?

Startup execution:
  loading initial environment

-->atomsInstall('SIVP')
 ans  =

!SIVP  0.5.3.2-1  allusers  SCI\contrib\SIVP\0.5.3.2-1  I  !

-->
```

Once the module is installed, you need to restart Scilab to make it work. As the module loads, messages are displayed in the console (see Figure 5.6).

Figure 5.6 : Loading of the SIVP module as Scilab starts up

```
File Edit Control Applications ?

Startup execution:
  loading initial environment

SIVP - Scilab Image and Video Processing Toolbox
        Load macros
        Load gateways
        Load help
        Load demos

-->
```

***Tip** › Certain external modules may need to be compiled during the installation process which may cause issues for Windows users. In this event, there are two solutions:*

- *Install a Microsoft Visual C++ redistribuable version which you can download directly from the **Microsoft** [http://d-booker.jo.my/sci-mvcpp] site.*
- *Install the Scilab MinGW supplementary module which was created for this explicit purpose.*

***Caution** › If you experience issues with a supplementary module, you can uninstall it manually by deleting its corresponding directory within the `SCI/contrib/` directory (see **Section 6.2, Installation**).*

6

Downloading and Installing Scilab

To conclude this presentation of the Scilab software, here is additional information that you may find useful during the Scilab download and installation process.

6.1. Where to find Scilab?

Scilab's official website [http://www.scilab.org/en] (see Figure 6.1) is without doubt the prime location to find all information on Scilab.

Figure 6.1 : Scilab's official website

A link on the homepage redirects to the download page [http://d-booker.jo.my/sci-download-en] which provides access to the latest Scilab versions for the three main operating systems (Windows, Linux et Mac OS).

Caution › *When downloading the Windows or Linux versions, you need to choose the version that best fits your processor's architecture:*

- *For a 32 bits Windows or Linux architecture, only the Scilab 32 bits version should be installed.*
- *For a Linux 64 bits architecture, install the Scilab 64 bits version.*
- *For Windows 64 bits, you can install either the 32 or 64 bits version. However, the Scilab 64 bits version is recommended to optimize the software performance.*
- *For MAC OS users, only one Scilab version is available.*

In Linux, you can also download and install Scilab through a package manager such as Synaptic.

Figure 6.2 : *Installing Scilab through Synaptic*

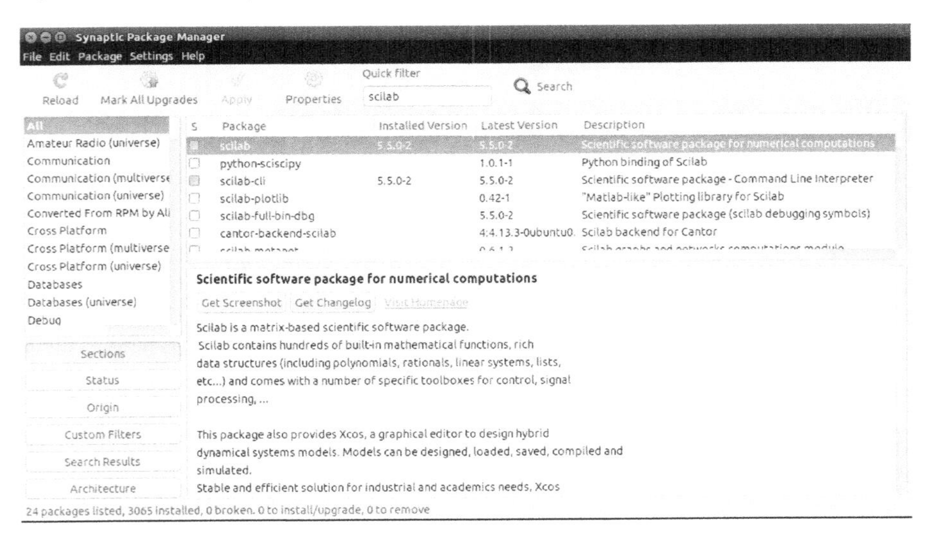

Caution › *Synaptic often does not offer the most up to date versions of Scilab (or of its associated packages). For this reason, it may be preferable to perform a manual installation.*

In some cases, you may also need to install different versions of Scilab, for example to get around certain bugs. On the same site, you can find:

- older version of Scilab which can be requested *via* a form [http://d-booker.jo.my/sci-previous-versions] (the list of Scilab versions [http://d-booker.jo.my/sci-history-versions] is available on the Scilab website)

- development versions, called *Nightly Builds* since they are compiled each night. They can be convenient to check if a recently reported bug was correctly fixed

  Caution › *The quality of these versions does not match the official version and should not be used in production!*

- sporadic *alpha* or *beta* versions to test new developments prior to major updates

Since 2008, Scilab has been open source under a license issued by CEA (Atomic Energy and Alternative Energies Commission), CNRS (the French National Centre for Scientific Research) and INRIA (the French Institute for Research in Computer Science and Automation), called CeCILL [http://d-booker.jo.my/cecill-en], which is also compatible with the GPL license [http://d-booker.jo.my/gnu-gpl]. In fact, Scilab's source code is accessible and modifiable! Scilab's development is managed through the Git tool [http://d-booker.jo.my/pro-git] that provides access to the source code which can be viewed by everyone and modified by contributors. The Scilab source code is accessible from http://cgit.scilab.org and http://gitweb.scilab.org (see Figure 6.3).

Figure 6.3 : *Accessing the Scilab source code from gitweb*

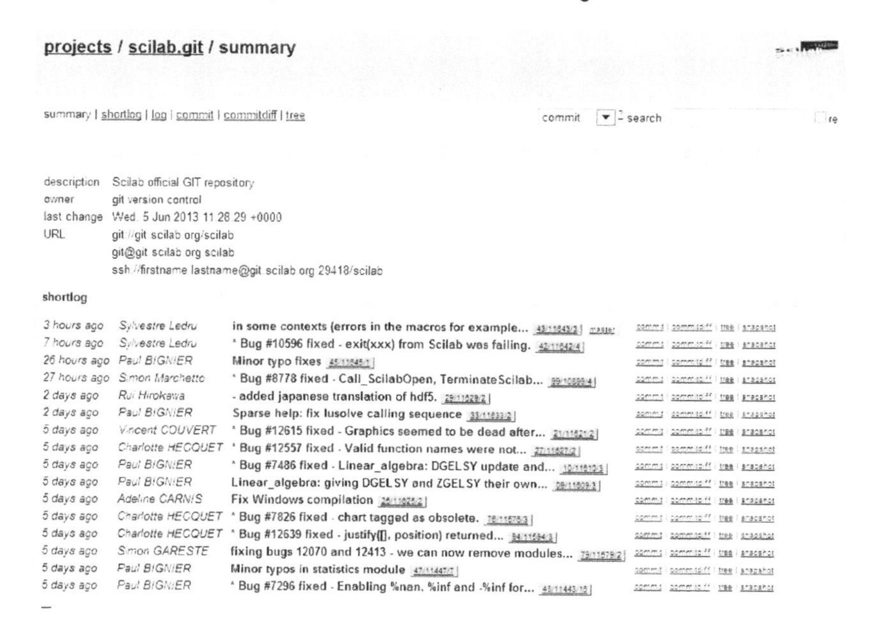

Figure 6.4 : Comparing versions on gitweb

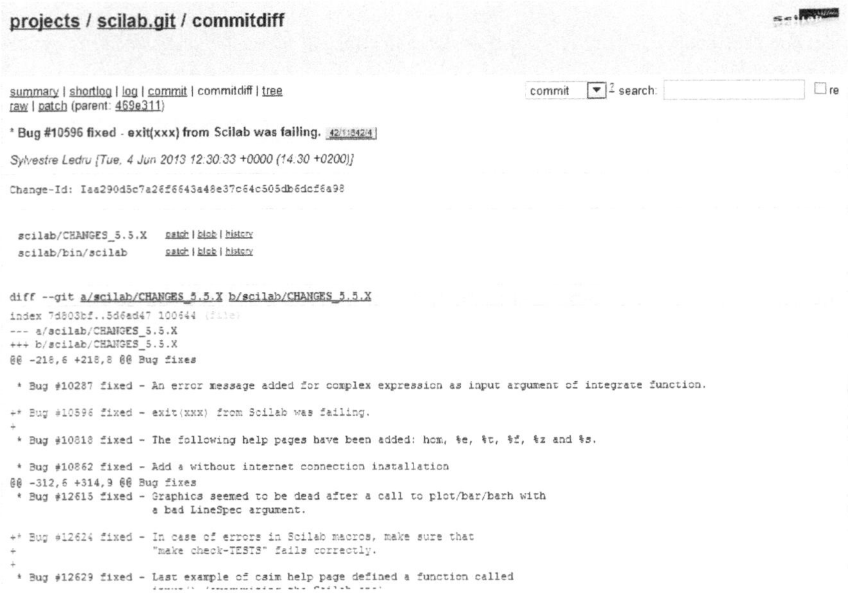

6.2. Installation

Once the appropriate binary file is downloaded, the installation is very simple. In accordance with your operating system:

- For Windows, execute the `scilab-*.exe` file, for example by double-clicking on the file icon from the file explorer. Then follow the instructions in the different windows (see Figure 6.5).

Figure 6.5 : Installing Scilab on Windows

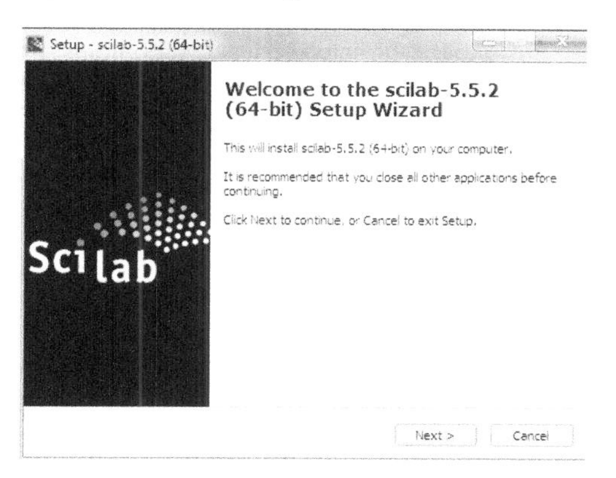

- In Linux, if you perform a manual installation, extract the `scilab-*.tar.gz` file in the specific directory where you wish to install Scilab. For example, execute the command `tar xvf scilab-5.5.2.tar.gz -C /usr/` from a terminal (you can also choose a directory other than `/usr/` and make sure you have administrative rights).
- On Mac OS, drag the `scilab-*.dmg` file icon (image of downloaded disk) to the Applications folder to launch the Mac OS applications installer (see Figure 6.6).

Figure 6.6 : Installing Scilab on Mac OS

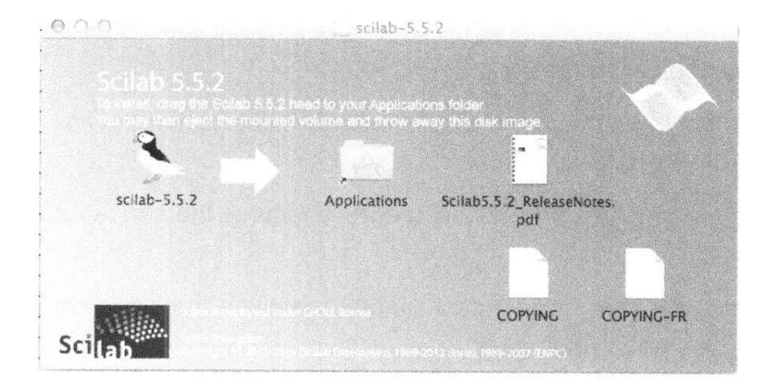

During the Scilab installation, and depending on your operating system, a group of files and programs are installed in a given directory. To find their location, follow the path:

- `C:/Program*/Scilab-*.*.*/` if you use Windows
- `/usr/Scilab-*.*.*/` if you use Linux

- `/Applications/Scilab.app/Contents/MacOS/share/scilab/` if you use Mac OS

From now on, we will refer to Scilab's installation directory as `SCI`. It contains, among other things, two important directories:

- `SCI/bin/`, which contains Scilab's main executables
- `SCI/contrib/` which in the future will contain supplementary modules

6.3. Executables and launch options

The executable file that is used to launch Scilab is located in the `SCI/bin/` directory. It is called:

- `scilab` in Linux or Mac OS
- `WScilex.exe` in Windows

If you wish to create shortcut to start Scilab, you need to point to one of these files. Once you launch the correct executable, Scilab's main window will appear as shown in Figure 2.1. You can also launch Scilab straight from your operating system's command-line interpreter by calling the executable file above. In this case, several launch options are available:

- `scilab -nwni`: launches Scilab in the terminal to use through the command-line without loading the graphics features.
- `scilab -nw`: launches Scilab in the terminal to use through the command-line while loading advanced features such as graphics.
- `scilab -e 'command'`: launches Scilab and silently executes the `command` Scilab during startup.
- `scilab -f file.sce`: launches Scilab and silently executes the commands file `file.sce` (also see Chapter *Scripts*).

Figure 6.7 : Launching Scilab from a command prompt

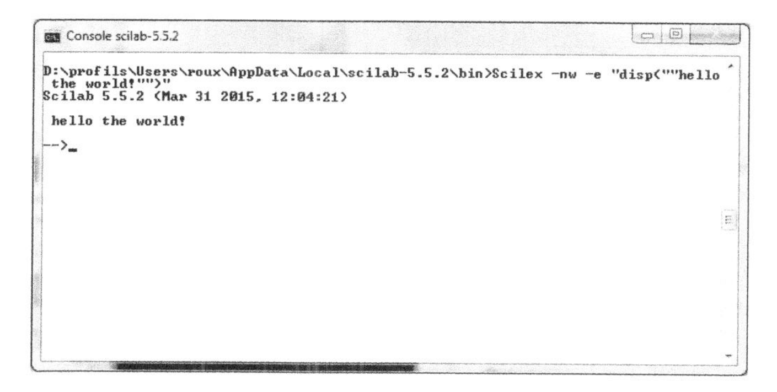

Tip › You can also find files located in the `SCI/bin/` *directory that let you execute Scilab directly from a terminal:*

- *On Windows,* `Scilex.exe` *performs the equivalent of launching Scilab with the option* `-nwni`.
- *On Linux or Mac OS:*
  - `scilab-cli` *corresponds to launching Scilab with the* `-nwni` *option*
  - `scilab-adv-cli` *corresponds to launching Scilab with the option* `-nw`.

There are also more advanced options:

- `scilab -ns` starts Scilab without the start file `scilab.start` (see Chapter *Scripts*).
- `scilab -nouserstartup` starts Scilab without loading the user start files `SCIHOME/.scilab` or `SCIHOME/scilab.ini` (see Chapter *Scripts*).
- `scilab -nb` starts Scilab without displaying the welcome banner.
- `scilab -l language` launches Scilab while configuring the user language, for example `"fr"` for french.
- `scilab -mem N` launches Scilab while setting the stack size N (memory allocated to Scilab).
- `scilab -version` displays a window with the Scilab version before returning control.

Caution › When launching Scilab from the command-line, the current directory is the one from which the command `scilab` *(or* `Wsilex`*) is called.*

Figure 6.8 : Launching Scilab with the -version option

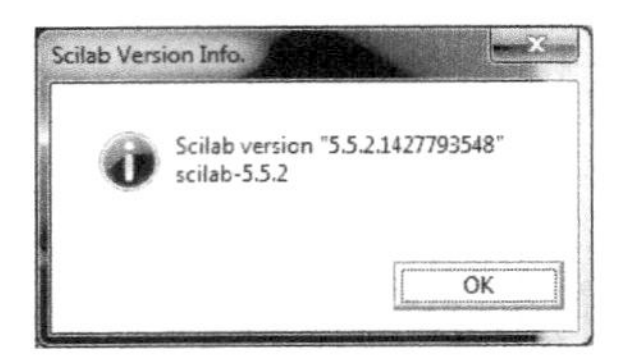

Tip › If Scilab's user interface language does not suit you, this parameter can be changed at any time from the preferences editor (see **Figure 2.3***). You can also retrieve and modify this data from the command-line with* **getlanguage** *and* **setlanguage**.

```
-->// retrieve the Scilab interface language

-->getlanguage() // in English at first
 ans  =
 en_US

-->// modify Scilab interface language

-->setlanguage('fr_FR')  // change to French
 ans  =
  T
```

Computing

The primary function of a numerical computation software such as Scilab is to easily perform all the mathematical calculations that an engineer may come across. This part will demonstrate how to perform these calculations, from manipulating real numbers, booleans, character strings, to complex objects such as matrices, polynomials and other lists.

7

Numbers and First Calculations

Scilab's primary attribute is its perfect adaptation to the mathematical syntax used to process numerical calculations.

7.1. Floating point numbers

Scilab's language is fundamentally based on floating point numbers. They allow the representation of decimal numbers following the IEEE754 standard [http://d-booker.jo.my/sci-ieee-floating]. These numbers are called real numbers with double precision since they occupy 64 bits of memory with:

- a sign s on one bit
- an exponent e over 11 bits
- a significand d over 52 bits

The corresponding number is $s \times d \times 2^{e-off}$ (with an offset off which depends on the size of the exponent). You can visualize the memory storage similarly to Figure 7.1 while keeping in mind that the significand's 52 bits roughly correspond to 16 significant digits for a decimal number.

Figure 7.1 : Scilab's real number representation

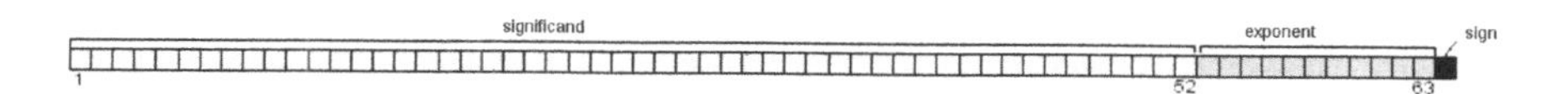

There are two ways of entering a value in the Scilab console, by using either of these notations:

- the decimal notation by using the period `.` as decimal mark
- the scientific notation by using `E` or `D` to denote the exponent

```
-->12.34     // decimal notation
 ans  =
    12.34

-->1.234D+1     // scientific notation
 ans  =
    12.34

-->1234E-2     // scientific notation
 ans  =
    12.34
```

Displaying a number defined by s, e, d in the console can be done, by default, with a maximum of nine characters, depending on the size of the number in decimal or scientific notation. However this behavior can be modified with the `format` command:

- `format('e',n)` displays a number in scientific format with `n-1` characters.
- `format('v',n)` displays a number following a variable format: it uses the decimal form if the number calls for less than `n-1` characters, otherwise uses the scientific form with `n-1` characters.

***Caution** › The value displayed by Scilab in the console is rounded and therefore does not reflect the exact memory content! (See **examples of rounding errors**.)*

```
-->x=1.2345678D+9
 x  =
    1234567800.

-->format('e',8),x     // scientific format with 7 characters
 ans  =
    1.2D+09

-->format('e',9),x     // scientific format with 8 characters
 ans  =
    1.23D+09

-->format('e',11),x     // scientific format with 10 characters
 ans  =
    1.2346D+09

-->format('e',12),x     // scientific format with 11 characters
 ans  =
    1.23457D+09

-->format('v',11),x     // variable format with 10 characters
 ans  =
    1.2346D+09

-->format('v',12),x     // variable format with 11 characters
 ans  =
```

```
    1234567800.

-->10*x
 ans  =
    1.23457D+10
```

7.2. Elementary mathematical functions

Operations on real numbers are coded in a very intuitive manner:

- addition +
- subtraction -
- multiplication *
- division /
- raising to a power ^ or **

You can use parentheses ( ), and the usual order of operations rules apply (* and / take precedence over + et -).

Caution › Make sure to use the period and not the comma as decimal separator! The comma is used as a command separator.

```
-->// order of operations and parentheses

-->1/2+3
 ans  =
    3.5

-->1/(2+3)
 ans  =
    0.2

-->2+3/10
 ans  =
    2.3

-->(2+3)/10
 ans  =
    0.5

-->// using parentheses with exponents

-->-567^12
 ans  =
  - 1.10407D+33

-->(-567)^12
 ans  =
```

```
     1.10407D+33

-->// decimal separator . and not ,

-->2+3.4
 ans  =
     5.4

-->2+3,4
 ans  =
     5.
 ans  =
     4.
```

Caution › When a calculation produces an error, such as a division by zero, Scilab stops the calculation at the location of the error. In some cases, you can modify this behavior with the **ieee** *function, which only displays a warning and lets Scilab carry on with the calculations.*

```
-->ieee(2),1/0     // infinite result
 ans  =
     Inf

-->ieee(1),1/0     // warning + infinite result
Warning : division by zero...

 ans  =
     Inf

-->ieee(0),1/0     // default behavior

     !--error 27

 Division by zero...
```

You can also modify the default response of Scilab straight from the preferences editor (menu EDIT/ PREFERENCES/GENERAL tab FLOATING POINT EXCEPTION, see figure **below**.*)*

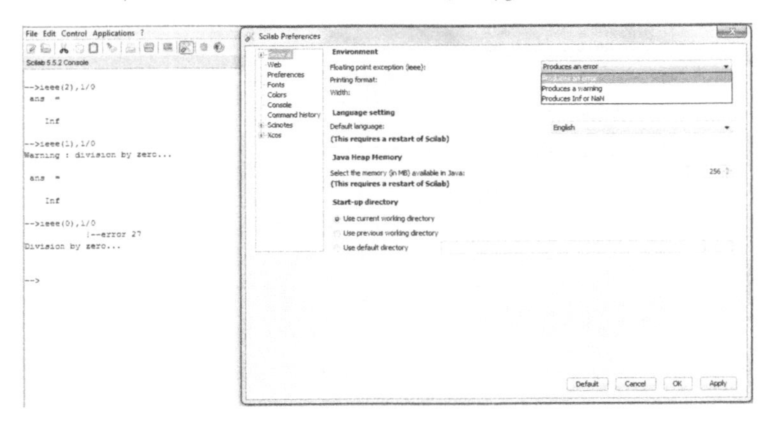

Functions of one variable

Elementary mathematical functions of a single real variable can be found in Scilab:

- the exponential `exp` and logarithmic functions (`log` the natural logarithm, or `log10` the logarithm with base 10, or `log2` the logarithm with base 2)
- the trigonometric functions (sine `sin`, cosine `cos`, tangent `tan` and their reciprocal functions `asin`, `acos`, `atan`)
- the hyperbolic functions `sinh`, `cosh`, `tanh` and their reciprocals `asinh`, `acosh`, `atanh`
- certain preprogrammed special functions such as `erf`, `gamma`, `legendre`, `besseli`
- the square root `sqrt`, absolute value `abs`, the sign function `sign`

They can be used by following the usual `function(argument)` syntax. You can combine these different functions within a single calculation by using standard binary/unary operators.

```
-->// nth roots

-->sqrt(3),3^(1/2)    // square roots
 ans  =
    1.732050808
 ans  =
    1.732050808

-->2^(1/3)    // cube root
 ans  =
    1.25992105

-->// exponent and logarithms

-->exp(log(2))
 ans  =
    2.

-->log(exp(2))
 ans  =
    2.

-->log10(1000)    // log base 10
 ans  =
    3.

-->// trigonometric functions

-->cos(1)^2+sin(1)^2
 ans  =
    1.

-->4*atan(1)    // =pi
```

```
 ans  =
    3.141592654
```

Functions of several variables

Functions of several variables are also at your disposal. Their syntax is of the form `function(argument1,argument2,...)`. For example:

- remainder of a Euclidean division with `pmodulo` or `modulo`
- perform prime factorization with `factor` or list all the prime numbers up to a certain limit with `primes`
- maximum and minimum with `max` and `min`

```
-->pmodulo(10,6)     // remainder of 10/6 within the [0;5] interval
 ans  =
    4.

-->max(1,2,3)     // maximum
 ans  =
    3.

-->min(2,1,3,4)     // minimum
 ans  =
    1.

-->primes(11)     // prime numbers up to 11
 ans  =
    2.    3.    5.    7.    11.

-->factor(12)     // break 12 down into prime factors
 ans  =
    2.    2.    3.
```

Caution › Unlike many other languages, Scilab does not have a `div` function that calculates the quotient of a Euclidean division $a = q \times b + r$. However, it is easy to see that the quotient q can be retrieved from `(a-pmodulo(a,b))/b` or by using rounding functions.

Common rounding functions

Scilab provides the following functions that perform rounding . The function `round` rounds to the nearest integer, `int` rounds towards zero, `ceil` rounds up, and `floor` rounds down.

```
-->floor(1.3),ceil(1.3),round(1.3),int(1.3)
 ans  =
    1.
 ans  =
    2.
 ans  =
```

```
     1.
 ans  =
     1.

-->floor(-1.3),ceil(-1.3),round(-1.3),int(-1.3)
 ans  =
  - 2.
 ans  =
  - 1.
 ans  =
  - 1.
 ans  =
  - 1.

-->floor(1.6),ceil(1.6),round(1.6),int(1.6)
 ans  =
     1.
 ans  =
     2.
 ans  =
     2.
 ans  =
     1.

-->floor(-1.6),ceil(-1.6),round(-1.6),int(-1.6)
 ans  =
  - 2.
 ans  =
  - 1.
 ans  =
  - 2.
 ans  =
  - 1.

-->// compute the quotient of a euclidean division by rounding

-->floor(10/6)    // quotient
 ans  =
     1.

-->10-6*floor(10/6)    // remainder
 ans  =
     4.

-->pmodulo(10,6)    // remainder
 ans  =
     4.

-->(10-pmodulo(10,6))/6    // quotient
 ans  =
     1.
```

Caution › *Watch out for rounding errors when working with double-precision real numbers! Their order of magnitude is given by the size of the significand which is approximately* 10^{-16}. *Specifically, calculations using integers greater than 16 digits are no longer accurate.*

```
-->// discrepancy between the display and result

-->sqrt(2)^2  // displays 2
 ans  =
    2.

-->sqrt(2)^2-2  // but the previous result is not 2
 ans  =
    4.44089D-16

-->// minor rounding error:

-->sqrt(6),sqrt(2)*sqrt(3)  // displays the same result
 ans  =
    2.449489743
 ans  =
    2.449489743

-->// however their results are slightly different:

-->sqrt(6)-sqrt(2)*sqrt(3)
 ans  =
  - 4.44089D-16

-->// significant rounding error:

-->X=1D30;Y=1D10;

-->X+Y-X  // result 0
 ans  =
    0.

-->X-X+Y  // result Y
 ans  =
    1.00000D+10

-->// complex rounding error:

-->%e^(%i*%pi)  // =-1 within a 10^(-16) error margin
 ans  =
  - 1. + 1.22465D-16i
```

If you do not wish to display calculation results in the console that are linked to rounding errors, you can use the `clean` function to round any result smaller than 10^{-10} to 0.

```
-->clean(10^(-9))        // not rounded
 ans  =
    1.00000D-09

-->clean(10^(-10))       // rounded to 0
 ans  =
    0.
```

```
-->%e^(%i*%pi)              // rounding error on complex part
 ans  =
  - 1. + 1.22465D-16i

-->clean(%e^(%i*%pi))  // error rounded to 0
 ans  =
  - 1.
```

On the other hand if you need to perform calculations with a very high precision, some supplementary modules provide that ability. To achieve a floating-point arithmetic precision of 10^{-32}, use the Double-Double [http://d-booker.jo.my/sci-dbldbl] module, which can be dowloaded *via* ATOMS. To perform calculations with multiple precisions, use the MPScilab [http://d-booker.jo.my/sci-mpscilab] module.

7.3. Integer formats

Although, by default, all integers are treated as double-precision real numbers, Scilab can also represent integers with a finite number of bits:

- signed integers with 8, 16 or 32 bits by using `int8`, `int16`, `int32`
- unsigned integers with 8, 16 or 32 bits by using `uint8`, `uint16`, `uint32`

The command `double` converts these integers to double-precision real numbers.

Caution › *Beware of erroneous results when adding signed and unsigned integers.*

```
-->2^4+2^7      // =144
 ans  =
    144.

-->uint8(2^4+2^7)      // =144
 ans  =
  144

-->uint8(2^4+2^7)+uint8(2^4+2^7)      // 144+144=32+256
 ans  =
  32

-->int8(2^4+2^7)      //=144-256
 ans  =
 -112

-->int8(2^4+2^7)+int8(2^4+2^7)      // =-112-112+256
 ans  =
  32
```

Scilab can also perform bitwise operations on binary representations of unsigned integers:

- AND with `bitand`
- OR with `bitor`
- complement with `bitcmp`

If you wish to better understand how these work, you can display the binary representation of integers with `dec2bin` and, conversely, with `bin2dec`.

```
-->x=uint8(6),dec2bin(x)    // 6=(0110)_2
 x  =
  6
 ans  =
 110

-->y=uint8(8),dec2bin(y)    // 8=(1000)_2
 y  =
  8
 ans  =
 1000

-->bitcmp(x,4),dec2bin(ans)    // 9=(1001)_2
 ans  =
  9
 ans  =
 1001

-->bitand(x,y),dec2bin(ans)    // 0=(0000)_2
 ans  =
  0
 ans  =
 0

-->bitor(x,y),dec2bin(ans)    // 14=(1110)_2
 ans  =
  14
 ans  =
 1110
```

8

Variables, Constants and Types

Variables make it possible to perform complex calculations which require multiple steps while storing intermediate results.

8.1. Creating variables

A Scilab variable is defined by its name (a chain of characters) which is assigned a value with the assignment operator `=`. The variable can then be used or modified in any calculation. The `clear` command erases it.

```
-->x=10      // creating a variable
 x  =
    10.

-->2*x+1     // calculation with x=10
 ans  =
    21.

-->x         // x is not modified
 ans  =
    10.

-->x=2*x+1   // modify x
 x  =
    21.

-->clear x   // erase x

-->x         // x no longer exists

    !--error 4

 Undefined variable: x
```

Caution › To erase operations displayed in the console, use the command `clc` *or* `tohome`*. Do not confuse it with* `clear`*, which erases the variables from memory.*

Every time a result is displayed in the console, the variable name it was stored in is displayed before the `=`. If it is not assigned a variable name, it gets stored in the variable `ans` (which then appears before the `=`).

Caution › *Creating a variable in Scilab is automatically achieved the moment it is assigned. There is no need to declare the variable's type before assigning it a value. This process is typical of script languages (such as bash, php, etc.), as opposed to compiled languages (such as C, FORTRAN, etc.).*

Nevertheless, each Scilab variable has a type which defines the kind of data it stores. You can retrieve this type as a character string with the command `typeof` or retrieve it as a number with `type`.

```
-->x=1     // floating-point real number
 x  =
    1.

-->typeof(x)     // type: "constant"
 ans  =
 constant

-->type(x)     // type # 1
 ans  =
    1.

-->y=int8(1)     // 8 bits integer
 y  =
  1

-->typeof(y)     // type: int8
 ans  =
 int8

-->type(y)     // type # 8
 ans  =
    8.

-->z=double(y)     // conversion of "integer" to "double"
 z  =
    1.

-->typeof(z)
 ans  =
 constant
```

Caution › *A variable's type can change at any time, depending on the value it is assigned!*

The name of a variable must obey certain rules:

- It may only contain lower or upper case letters without accents, numbers or the following characters: `% _ # ! $ ?`.
- It cannot begin with a number.
- Only the first 24 characters are recorded by Scilab.

Variable names that are too long are automatically truncated at 24 characters (you are given a warning).

```
-->a_variable_name_that_is_too_long=1
Warning :
The identifier : a_variable_name_that_is_too_long
 has been truncated to: a_variable_name_that_is_.

 a_variable_name_that_is_  =
    1.

-->a_variable_name_that_is_too_long // contains the value 1
Warning :
The identifier : a_variable_name_that_is_too_long
 has been truncated to: a_variable_name_that_is_.

 ans  =
    1.

-->a_variable_name_that_is_the_same=2
Warning :
The identifier : a_variable_name_that_is_the_same
 has been truncated to: a_variable_name_that_is_.

 a_variable_name_that_is_  =
    2.

-->a_variable_name_that_is_the_same // contains the value 2
Warning :
The identifier : a_variable_name_that_is_the_same
 has been truncated to: a_variable_name_that_is_.

 ans  =
    2.

-->a_variable_name_that_is_too_long // also contains 2
Warning :
The identifier : a_variable_name_that_is_too_long
 has been truncated to: a_variable_name_that_is_.

 ans  =
    2.

-->a_variable_name_that_is_ // identifier for both variables
 ans  =
    2.
```

Caution › *Remember that variable names cannot contain:*

- *accented characters: àâéèêëïöùü...*
- *special characters that symbolize an operation:* + - * / ^ ' & | ~ . = < > , ; :
- *special characters used as brackets* [ ], ( ), { }

8.2. Mathematical constants

In Scilab, there are certain special variables whose value cannot be modified. Their name always begins with the character `%` (percent). You can list them with the command `predef("names")`. The list contains:

- the mathematical constants `%pi`, `%e` (base of the natural logarithm) and the complex number `%i`
- computer constants such as the machine epsilon `%eps` or for results that cannot be computed `%nan` or `%inf`
- the booleans True `%T` (or `%t`) and False `%F` (or `%f`) (see Chapter *Booleans*)

```
-->// mathematical constants

-->a=cos(%pi)
 a  =
  - 1.

-->b=log(%e)
 b  =
    1.

-->// complex numbers

-->%i^2     // =-1
 ans  =
  - 1.

-->(1+%i)^2     // =2i
 ans  =
    2.i

-->c=exp(%i*%pi)     // =-1 (within rounding error!)
 c  =
  - 1. + 1.22465D-16i

-->ans     // contains the result of 2*%i
 ans  =
    2.i

-->// NaN, infinity and machine epsilon

-->typeof(%eps)     // machine epsilon
 ans  =
 constant

-->x=10^155,x^2     // >10^309 yields infinity
 x  =
    1.0000D+155
 ans  =
```

```
    Inf

-->%inf-%inf,%inf/%inf      // not defined, returns %nan
 ans  =
    Nan
 ans  =
    Nan
```

The number `%i` lets you manipulate complex numbers in Scilab. Incidentally, several of the aforementioned functions accept complex numbers as inputs. The following functions are also available:

- `conj` to retrieve the complex conjugate
- `real` and `imag` to return the real and imaginary part
- `polar` returns the polar representation of a complex number

```
-->z=1+2*%i     // a complex number
 z  =
    1. + 2.i

-->conj(z)     // complex conjugate
 ans  =
    1. - 2.i

-->real(z)     // real part
 ans  =
    1.

-->imag(z)     // imaginary part
 ans  =
    2.

-->[r,a]=polar(z)     // polar representation
 a  =
    1.107148718 + 5.55112D-17i
 r  =
    2.236067977

-->r*exp(%i*a)     // =z
 ans  =
    1. + 2.i

-->sqrt(-1)     // principal sqrt of -1
 ans  =
    i
```

Caution › *In Scilab, there is no difference between real numbers and complex numbers. Both cases belong to the* constant *type. Certain functions sometimes even return complex values when given real inputs.*

From a calculation point of view, the constant `%inf` behaves like the mathematics infinity ∞. This means Scilab is able to avoid certain calculation errors, such as errors linked to memory excess, by replacing the result with `%inf` whenever possible.

```
-->1+%inf
 ans  =
    Inf

-->-2*%inf
 ans  =
  - Inf

-->1/%inf
 ans  =
    0.

-->%inf+%inf
 ans  =
    Inf

-->%inf*%inf
 ans  =
    Inf

-->ieee(2)    // ieee exception mode

-->1/0    // returns infinity
 ans  =
    Inf
```

8.3. Advanced variable management

The `save` command lets you save variables in a binary file with extension `.sod` (which stands for *Scilab Open Data*). You can then load these variables again into Scilab with the command `load`.

```
-->x=%e*%pi,y=%i, // two variables
 x  =
    8.539734223
 y  =
    i

-->save('myvar.sod','x','y') // saving to the file myvar.sod

-->ls('*.sod')    // file created in the current directory
 ans  =
 myvar.sod

-->clear x    // erase x
```

```
-->y=y+1    // modify y
 y  =
    1. + i

-->load('myvar.sod')    // reload x and y

-->x,y
 x  =
    8.539734223
 y  =
    i
```

If you created a large number of variables, you can list them with the commands `who`, `whos` and `who_user`.

```
-->x=%e*%pi    // variable x
 x  =
    8.5397342

-->who    // list all variables

Your variables are:

                    x                         id                     filename
                 path                    fichier                         whos
development_toolslib              scicos_autolib            scicos_utilslib
         graphicslib                datatipslib                        guilib
         scinoteslib                      jnull                         jvoid
      preferenceslib              helptoolslib                    tclscilib
         atomsguilib                   matiolib                parameterslib
          umfpacklib             spreadsheetlib               demo_toolslib
          randliblib       external_objectslib                        enull
            soundlib                  m2scilib  compatibility_functilib
       statisticslib            windows_toolslib                         WSCI
           stringlib      special_functionslib                    sparselib
                   %z                         %s              polynomialslib
     optimsimplexlib              optimbaselib                neldermeadlib
    interpolationlib          linear_algebralib             output_streamlib
          integerlib             dynamic_linklib         data_structureslib
           fileiolib              functionslib                          SCI
             corelib                        PWD                          %tk
                   %T                       %nan                        %inf
             SCIHOME                     TMPDIR                         %gui
                    $                         %t                           %f
                  %io                         %i                          %pi


 using       17871 elements out of    10000000.
 and           100 variables out of       9231.

Your global variables are:

     %modalWarning                %toolboxes              %toolboxes_dir
     %helps
```

```
 using          39 elements out of      999999.
 and             4 variables out of        767.

-->who_user()    // user variables
 User variables are:

 x            id          filename            path
 fichier      whos

 Using 8581 elements out of 9990093
 ans  =
!x            !
!             !
!id           !
!             !
!filename     !
!             !
!fichiertxt   !
!             !
!path         !
!             !
!fichier      !
!             !
!whos         !
!             !
!path         !
!             !
!help         !


-->whos -type constant     // variables of type "constant"
Name                         Type           Size           Bytes

%helps                       constant*      0 by 0         16
%i                           constant       1 by 1         32
%inf                         constant       1 by 1         24
%io                          constant       1 by 2         32
%nan                         constant       1 by 1         24
%pi                          constant       1 by 1         24
%toolboxes                   constant*      0 by 0         16
id                           constant       1 by 1         24
x                            constant       1 by 1         24

-->whos -name %     // variables containing the character %
Name                         Type           Size           Bytes

%
%gui                         boolean        1 by 1         24
%helps                       constant*      0 by 0         16
%i                           constant       1 by 1         32
%inf                         constant       1 by 1         24
%io                          constant       1 by 2         32
%modalWarning                boolean*       1 by 1         24
%nan                         constant       1 by 1         24
```

```
    %pi                     constant        1 by 1          24
    %s                      polynomial      1 by 1          56
    %T                      boolean         1 by 1          24
    %t                      boolean         1 by 1          24
    %tk                     boolean         1 by 1          16
    %toolboxes              constant*       0 by 0          16
    %toolboxes_dir          string*         1 by 1          248
    %z                      polynomial      1 by 1          56

    -->editvar("x")     // editing the variable x
```

Recall that Scilab has a variable browser and variable editor if you wish to perform certain operations from a graphics window rather than use the console.

Figure 8.1 : *Variable editor*

```
Startup execution:
  loading initial environment

-->x=%e*%pi
 x  =

    8.5397342

-->whos -type constant
Name                 Type           Size           Bytes

%e                   constant       1 by 1         24
%eps                 constant       1 by 1         24
%i                   constant       1 by 1         32
%inf                 constant       1 by 1         24
%io                  constant       1 by 2         32
%nan                 constant       1 by 1         24
%pi                  constant       1 by 1         24
%toolboxes           constant*      0 by 0         16
x                    constant       1 by 1         24

-->editvar("x")

-->
```

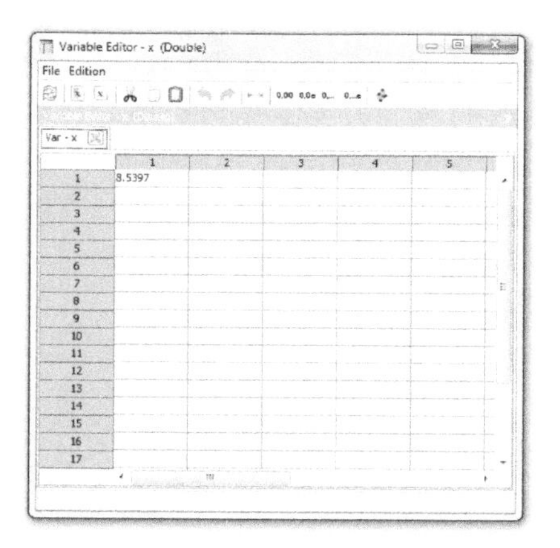

Caution › *The scope of variables in Scilab, i.e. the difference between a local variable and a global variable, will be discussed in* Part Programming.

The amount of memory available for variable storage is called the *stack*. You can find out the amount of memory used by Scilab and modify the amount of available memory with the functions `stacksize` and `gstacksize`. The stack used by Scilab is not only limited by the amount of memory available on your device but also by the number of pointers available to index stack elements. For a 32 bits architecture, a maximum of 2^{32} memory addresses are possible. This theoretically limits the amount of available memory to 4 Gb. With Windows, for technical reasons, the actual memory limit is 2 Gb. If one takes into account the amount of memory used to load various Scilab libraries, then only around 1 Gb remains. For 64 bits versions, this limit doubles to 2 Gb. To have access to as much memory as possible, use the command `stacksize('max')`.

```
-->// scilab 32bits
-->stacksize('max')

-->a=stacksize();

-->a(1)*8  // memeory space in bits
 ans  =

    1.001D+09

-->// scilab 64bits
-->stacksize('max')

-->a=stacksize();

-->a(1)*8 // memory space in bits
 ans  =

    1.792D+09
```

Caution › *Starting from Scilab's version 6, the stack size limit will be removed.*

9

Matrices

Matrices are the most important and most frequently used objects in Scilab. This type is used to portray tables with one or two dimensions in a unified way.

9.1. Creating and modifying

There are several ways of creating tables in Scilab. The most simple is based on the concatenation operator `[]`. To create a matrix, enter a list of values, row by row, in between brackets while obeying the following syntax rules:

- Values on the same row are separated by spaces or a comma (,).
- Rows are separated by semicolons (;).

The console commands can be entered over several lines (as long as you don't forget to close the matrix correctly).

Caution › *Contrary to other languages, the cell/row/column numbering in Scilab starts at 1 (rather than at 0).*

The command `size` returns the size of a matrix (number of rows and columns). The result of this command is a matrix with one row and two columns.

```
-->A=[1 2 3; 4 5 6; 7 8 9] // matrix 3 lines and 3 columns
 A  =
    1.    2.    3.
    4.    5.    6.
    7.    8.    9.

-->typeof(A) // same type as real numbers
 ans  =
 constant

-->size(A) // size of A
 ans  =
    3.    3.

-->B=[10,11,12;15 14 13] // matrix with 2 rows and 3 columns
 B  =
    10.    11.    12.
```

```
    15.    14.    13.

-->size(B) // size of B
 ans  =
    2.    3.

-->// enter over multiple lines

-->[1 2 3;
-->3 4 5]
 ans  =
    1.    2.    3.
    3.    4.    5.
```

Tip › The [ ] command produces an empty matrix, with zero rows and zero columns. A number (real, complex, integer) is considered a matrix with one line and one column.

Matrices can later be concatenated to create larger ones, by using `[ ]`, as long as they have the right dimensions.

```
-->A=[1 2 3; 4 5 6; 7 8 9];

-->B=[10,11,12;15 14 13];

-->C=[A;B]       // A above B
 C  =
    1.     2.     3.
    4.     5.     6.
    7.     8.     9.
    10.    11.    12.
    15.    14.    13.

-->D=[B;A]       // B above A
 D  =
    10.    11.    12.
    15.    14.    13.
    1.     2.     3.
    4.     5.     6.
    7.     8.     9.

-->E=[C,D]       // C left of D
 E  =
    1.     2.     3.     10.    11.    12.
    4.     5.     6.     15.    14.    13.
    7.     8.     9.     1.     2.     3.
    10.    11.    12.    4.     5.     6.
    15.    14.    13.    7.     8.     9.

-->F=[D,C]       // D left of C
 F  =
    10.    11.    12.    1.     2.     3.
    15.    14.    13.    4.     5.     6.
    1.     2.     3.     7.     8.     9.
    4.     5.     6.     10.    11.    12.
```

```
    7.      8.      9.      15.     14.     13.

-->G=[A,B]       // A left of B -> error

    !--error 5

 Inconsistent column/row dimensions.
```

You can also concatenate matrices with the `cat` command.

```
-->A=[1 2 3; 4 5 6; 7 8 9]    // 3x3 matrix (3 rows, 3 columns)
 A  =
    1.    2.    3.
    4.    5.    6.
    7.    8.    9.

-->B=[10,11,12;15 14 13]    // 2x3 matrix (2 rows, 3 columns)
 B  =
    10.    11.    12.
    15.    14.    13.

-->// concatenating matrices

-->C=[A;B]     // A above B
 C  =
    1.     2.     3.
    4.     5.     6.
    7.     8.     9.
    10.    11.    12.
    15.    14.    13.

-->cat(1,A,B)     // =C
 ans  =
    1.     2.     3.
    4.     5.     6.
    7.     8.     9.
    10.    11.    12.
    15.    14.    13.

-->D=[B;A]     // B above A
 D  =
    10.    11.    12.
    15.    14.    13.
    1.     2.     3.
    4.     5.     6.
    7.     8.     9.

-->cat(1,B,A)     // =D
 ans  =
    10.    11.    12.
    15.    14.    13.
    1.     2.     3.
    4.     5.     6.
    7.     8.     9.
```

```
-->E=[C,D]      // C left of D
 E  =
    1.      2.      3.      10.     11.     12.
    4.      5.      6.      15.     14.     13.
    7.      8.      9.      1.      2.      3.
    10.     11.     12.     4.      5.      6.
    15.     14.     13.     7.      8.      9.

-->cat(2,C,D)     // =E
 ans  =
    1.      2.      3.      10.     11.     12.
    4.      5.      6.      15.     14.     13.
    7.      8.      9.      1.      2.      3.
    10.     11.     12.     4.      5.      6.
    15.     14.     13.     7.      8.      9.
```

Caution › *The matrices you wish to concatenate need to have compatible dimension, otherwise you will get an error message (5 or 6) stating* Inconsistent row/column dimensions.

You can generate certain special matrices automatically by providing their size as input to the following functions:

- `zeros` produces a zero matrix and `ones` yields a matrix full of 1s.
- `eye` creates the identity matrix (with 1s on the main diagonal and 0s elsewhere).
- `rand` fills a matrix with pseudorandom numbers uniformly distributed in the interval $[0;1[$ (also see `grand` which is used to generate random numbers according to different distributions as described online `help grand`).

```
-->O=zeros(2,3)     // zero matrix
 O  =
    0.    0.    0.
    0.    0.    0.

-->U=ones(4,3)     // matrix of ones
 U  =
    1.    1.    1.
    1.    1.    1.
    1.    1.    1.
    1.    1.    1.

-->Id=eye(3,3)     // identity matrix
 Id  =
    1.    0.    0.
    0.    1.    0.
    0.    0.    1.

-->M=rand(2,2)     // random numbers
 M  =
    0.537622980    0.225630349
    0.119992550    0.627409308
```

*Tip › When the functions **zeros**, **ones**, **eye** and **rand** take a matrix as argument, they generate a matrix of the same size as the input matrix. More specifically, if the input is a number, these functions create a 1#1 matrix (i.e. just a number!).*

```
-->A=rand(2,2)
 A  =
    0.280649802     0.778312860
    0.128005846     0.211903045

-->// creates a zero matrix of the same size as A

-->zeros(A)
 ans  =
    0.    0.
    0.    0.

-->// zeros(2) does not create a vector with two entries

-->zeros(2)
 ans  =
    0.
```

To retrieve the different values stored inside a matrix, specify the row and column of the entry of interest in between parentheses ( ). To modify a matrix value, use the = to assign a new value to the entry.

```
-->A=[1 2 3; 4 5 6; 7 8 9]     // 3x3 matrix (3 rows, 3 columns)
 A  =
    1.     2.     3.
    4.     5.     6.
    7.     8.     9.

-->A(2,3)     // value stored at row 2, column 3
 ans  =
    6.

-->A(2,3)=-1     // modify the value at row 2 column 3
 A  =
    1.     2.     3.
    4.     5.   - 1.
    7.     8.     9.

-->A(4,5)=10     // this assignment increases the size of A
 A  =
    1.     2.     3.     0.     0.
    4.     5.   - 1.     0.     0.
    7.     8.     9.     0.     0.
    0.     0.     0.     0.     10.

-->     // entry that doesn't exist in A

-->A(10,10)     // this call returns error 21
```

```
    !--error 21
Invalid index.
```

Tip › If you assign a value to a matrix entry that does not exist (the row/column number is greater than the matrix size), Scilab automatically increases the size of the matrix to assign the new value. It also incidentally fills the other entries that were created in the process with 0s.

You can also modify matrix entries via a graphical interface with the function `x_matrix` (see Figure 9.1) or by using the variable editor with the command `editvar` (see Figure 8.1 or Figure 9.1).

***Figure 9.1** : Graphical interface to modify matrix entries*

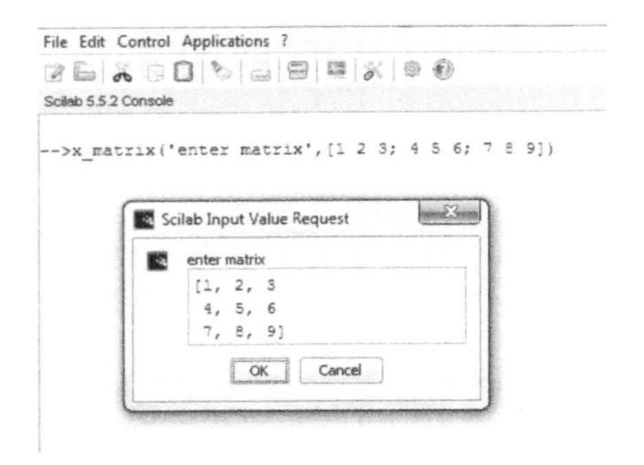

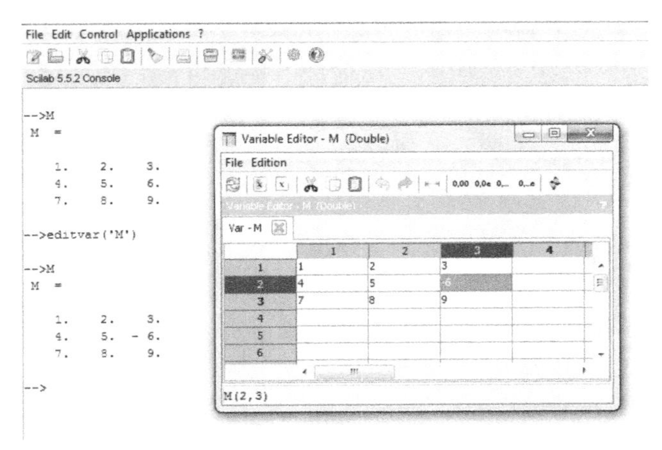

Caution › If you try and access a matrix value with an index that exceeds the matrix size, Scilab returns an execution error which states Invalid index. *This is the case when calling the entry/row/column number 0 (or indexing with negative or non-integer numbers).*

Matrices that have only one row or column (which are commonly called vectors) do not belong to a particular type in Scilab. However, Scilab allows the use of a simplified syntax to manipulate them. A vector entry can be accessed by specifying *one index rather than two*. To calculate the size of a vector, use the command `length` which computes the length of the vector. This command is better suited to vectors than the command `size`.

```
-->u=[1 2 3 4 5]    // row vector
 u  =
    1.    2.    3.    4.    5.

-->v=[-1;-2;-3]    // column vector
 v  =
  - 1.
  - 2.
  - 3.

-->    // size and length

-->size(u),length(u)
 ans  =
    1.    5.
 ans  =
    5.

-->size(v),length(v)
 ans  =
    3.    1.
 ans  =
    3.

-->// retrieving values

-->u(1,4),u(4)
 ans  =
    4.
 ans  =
    4.

-->v(2,1),v(2)
 ans  =
  - 2.
 ans  =
  - 2.
```

Caution › *To differentiate between vectors and matrices, you can use the function* **`isvector`**. *On the other hand, the command* **`ndims`**, *which returns the number of dimensions of a table, returns the value 2 for both matrices and vectors!*

```
-->M=eye(2,2);    // a matrix

-->ndims(M)    // 2 dimensions
 ans  =
```

```
      2.

-->u=[1:3];      // a vector

-->ndims(u)      // 1 dimension!
 ans  =
      2.

-->isvector(u)      // true
 ans  =
  T

-->isvector(M)      // false
 ans  =
  F
```

If you access a value of a matrix with more than one row and column and only provide one index, rather than the row and column number, Scilab treats the matrix as a vector obtained by concatenating the different matrix columns. In other words, Scilab behaves as if the matrix values were stored in memory sequentially, column by column. In this representation, the entry index is comparable to the numbering of a board game squares.

```
-->A=[1 2 3;4 5 6;7 8 9]
 A  =
    1.    2.    3.
    4.    5.    6.
    7.    8.    9.

-->A(4)      // coefficient A(1,2)
 ans  =
    2.

-->A(8)      // coefficient A(2,3)
 ans  =
    6.
```

*Tip › A matrix **A** can be transformed into a vector **u** by concatenating columns with the command **u=A(:)**.*

```
-->A=[1 2; 3 4]      // matrix with 2 rows 2 columns
 A  =
    1.    2.
    3.    4.

-->A(:)      // vector  (column)
 ans  =
    1.
    3.
    2.
    4.

-->//application
```

```
-->B=zeros(5,5);      // create a square matrix

-->B(:)=[1:25]      // fill matrix with values in one line
 B  =
     1.     6.      11.     16.     21.
     2.     7.      12.     17.     22.
     3.     8.      13.     18.     23.
     4.     9.      14.     19.     24.
     5.     10.     15.     20.     25.
```

The increment operator (:) can be used to create vectors from a list of equally spaced values. The general syntax to create a vector of values ranging from a to b in increments of step is a:step:b. If the increment is 1, then the syntax can be simplified to a:b. You can create the same types of vectors by using the command linspace by specifying an interval and the desired number of values (also see logspace, for values with logarithmic spacing, by searching the online help help logspace).

```
-->L=[1:5]     // values from 1 to 5
 L  =
    1.    2.    3.    4.    5.

-->U=[1:2:10]     // values from 1 to 10 (increment by 2)
 U  =
    1.    3.    5.    7.    9.

-->V=[5:-1:1]     // negative step
 V  =
    5.    4.    3.    2.    1.

-->linspace(1,5,3)     // 3 values equally spaced
 ans  =
    1.    3.    5.

-->linspace(1,5,5)     // 5 values equally spaced
 ans  =
    1.    2.    3.    4.    5.

-->logspace(-2,6,5)     // logarithmic spacing
 ans  =
    0.01    1.    100.    10000.    1000000.
```

Tip › *Here are some very useful shortcuts to retrieve matrix values: you can extract part of a matrix* **A** *(e.g. several entries over several rows/columns) with commands of the type* **A(X,Y)**, *where* **X** *and* **Y** *are vectors which specify the rows and columns of interest. This command can be used together with the* **:** *operator in order to retrieve part of a matrix* **A** *located within the rows* **a** *and* **b** *and the columns* **c** *and* **d** *by entering* **A(a:b,c:d)**. *Following this syntax,* **b** *and* **d** *can be replaced with the symbol* **$** *to designate the index of the last row or column. If* **:** *is used without values before or after, then all the rows (or columns) are called.*

```
-->A=[1 2 3; 4 5 6; 7 8 9]; // 3x3 matrix
```

```
-->A(4,5)=10 // this assignment increases the size of A
 A  =
    1.    2.    3.    0.    0.
    4.    5.    6.    0.    0.
    7.    8.    9.    0.    0.
    0.    0.    0.    0.    10.

-->A(:,4)=[-1;-2;-3;-4] // assign values to column 4
 A  =
    1.    2.    3.  - 1.    0.
    4.    5.    6.  - 2.    0.
    7.    8.    9.  - 3.    0.
    0.    0.    0.  - 4.    10.

-->A(2:3,3:4) // retrieve rows 2,3 and columns 3,4
 ans  =
    6.  - 2.
    9.  - 3.

-->A([2,4],[1,5]) // retrieve rows 2,4 and columns 1,5
 ans  =
    4.    0.
    0.    10.

-->A(1,1:$-1) // extract line 1 the first to the second-to-last column
 ans  =
    1.    2.    3.  - 1.
```

Vectors can be used to easily represent queues and stacks in Scilab. If we define `L` as a Scilab row vector, then:

- `L=[]` creates an empty stack/queue.
- `L=[x,L]` adds `x` to the start of the stack.
- `L=[L,x]` adds `x` to the end of the queue.
- `x=L(1)` retrieves the value of the stack/queue's first entry.
- `L(1)=[]` removes the stack/queue's first entry.

```
-->// stack representation

-->L=[]     // empty stack
 L  =
     []

-->L=[1,L]     // add 1 to the stack
 L  =
    1.

-->L=[2,L]     // add 2 to the stack
 L  =
    2.    1.
```

```
-->L=[3,L]    // add 3 to the stack
 L  =
    3.    2.    1.

-->x=L(1),L(1)=[]    // "unstacking"
 x  =
    3.
 L  =
    2.    1.

-->// queue representation

-->F=[]    // empty queue
 F  =
     []

-->F=[F,1]    // add 1 to the queue
 F  =
    1.

-->F=[F,2]    // add 2 to the queue
 F  =
    1.    2.

-->F=[F,3]    //add 3 to the queue
 F  =
    1.    2.    3.

-->x=F(1),F(1)=[]    // remove from the queue
 x  =
    1.
 F  =
    2.    3.
```

A disadvantage of matrices is that they occupy a large amount of memory space, even if a large number of entries are null. To remedy this problem, Scilab has a specific type called *sparse matrices*. These matrices can be manipulated by following the same syntax as regular matrices. To create a sparse matrix, use the specialized functions `spzeros`, `spones`, `speye`, `sprand` or transform a regular matrix into a sparse matrix by using the command `sparse`. You can also convert sparse matrices into regular matrices with the command `full`.

```
-->A=eye(5,5)    // identity matrix
 A  =
    1.    0.    0.    0.    0.
    0.    1.    0.    0.    0.
    0.    0.    1.    0.    0.
    0.    0.    0.    1.    0.
    0.    0.    0.    0.    1.

-->typeof(A)    // constant type
```

```
 ans  =
 constant

-->A=sparse(A)     // converts A into a sparse matrix
 A  =
(    5,    5) sparse matrix

(    1,    1)        1.
(    2,    2)        1.
(    3,    3)        1.
(    4,    4)        1.
(    5,    5)        1.

-->typeof(A)     // sparse type
 ans  =
 sparse

-->A(3,2)=1     // sparse matrix value assignment
 A  =
(    5,    5) sparse matrix

(    1,    1)        1.
(    2,    2)        1.
(    3,    2)        1.
(    3,    3)        1.
(    4,    4)        1.
(    5,    5)        1.

-->full(A)     // convert back to a "regular" matrix
 ans  =
    1.    0.    0.    0.    0.
    0.    1.    0.    0.    0.
    0.    1.    1.    0.    0.
    0.    0.    0.    1.    0.
    0.    0.    0.    0.    1.

-->sprand(5,5,0.2)     // "random" sparse matrix
 ans  =
(    5,    5) sparse matrix

(    2,    1)        0.097131776
(    3,    4)        0.823457824
(    3,    5)        0.821903265
(    5,    2)        0.994068494
```

9.2. Element-wise and matrix operations

Real number operations naturally also apply to matrices and are called element-wise operations. Given an operation on real numbers, Δ the corresponding element-wise operation is defined by the following formula:

$$C = A \,\Delta\, B \;\Leftrightarrow\; \forall\, i, j,\;\; C_{i,j} = A_{i,j} \,\Delta\, B_{i,j}$$

where Δ acts on matrices of the same size. The six element-wise operations are:

- addition +
- subtraction -
- multiplication .*
- right array division ./ or left array division .\
- raising to a power .^

```
-->A=zeros(3,3);A(:)=[1:9]
 A  =
    1.    4.    7.
    2.    5.    8.
    3.    6.    9.

-->B=2-eye(3,3)
 B  =
    1.    2.    2.
    2.    1.    2.
    2.    2.    1.

-->A+B // addition
 ans  =
    2.    6.    9.
    4.    6.    10.
    5.    8.    10.

-->A-B // subtraction
 ans  =
    0.    2.    5.
    0.    4.    6.
    1.    4.    8.

-->A.*B // multiplication
 ans  =
    1.    8.     14.
    4.    5.     16.
    6.    12.    9.

-->B./A // right division
 ans  =
    1.             0.5            0.285714286
    1.             0.2            0.25
    0.666666667    0.333333333    0.111111111

-->B.\A // left division
 ans  =
    1.     2.    3.5
    1.     5.    4.
    1.5    3.    9.

-->A.^B // raising to a power
```

```
 ans  =
    1.     16.     49.
    4.     5.      64.
    9.     36.     9.

-->A.^2 // raising to a power
 ans  =
    1.     16.     49.
    4.     25.     64.
    9.     36.     81.
```

Caution › *If* x *is a number and* **A** *a matrix, then* x+**A** *is equivalent to performing and element-wise addition of* **A** *and the matrix of the same size as* **A**: $\begin{pmatrix} x & \dots & x \\ \vdots & & \vdots \\ x & \dots & x \end{pmatrix}$. *Likewise for the operations* x-**A**, x*A, x^**A**.

There also exists a transpose operator .', which swaps the rows and columns and is formally written as:

$$B = A' \Rightarrow \forall\, i, j,\ B_{i,j} = A_{j,i}$$

You can also perform more complex permutations of rows and columns with the function permute.

Caution › *Do not confuse the transpose operator* .' *with the operator* ' *which computes the adjoint of a matrix, in other words the complex conjugate of the matrix transpose.*

Note that the pertrans command performs an operation similar to ': it computes the complex conjugate of the symmetric of a matrix with the symmetry performed about the second diagonal.

```
-->A=[1,%i;2,0;3,1+%i]     // a matrix
 A  =
    1.     i
    2.     0
    3.     1. + i

-->A.'     // transpose of A
 ans  =
    1.      2.     3.
    i       0      1. + i

-->permute(A,[2,1])     // =A.'
 ans  =
    1.      2.     3.
    i       0      1. + i

-->A'    // adjoint of A
 ans  =
    1.      2.     3.
  - i       0      1. - i
```

```
-->pertrans(A)     // adjoint with respect to the second diagonal
 ans  =
    1. - i      0    - i
    3.          2.     1.
```

In addition to element-wise operations, there are also more complex matrix operations. These operations are related to the use of matrices to represent linear maps. For example:

- Matrix multiplication is denoted the same way as the multiplication of real numbers `*` and is mathematically defined by:

  $$C = A * B \Leftrightarrow \forall\, i, j,\ \ C_{i,j} = \sum_k A_{i,k} * B_{k,j}$$

  This multiplication is not defined for matrices with incompatible sizes (the number of columns of A must be equal to the number of rows of B).
- Raising a matrix to a power indicated with `^` in Scilab is formally defined as:

  $$A^n = \underbrace{A \times A \times \cdots \times A}_{n \text{ matrix multiplication}}$$

  and is only well defined for an integer n.
- Left matrix division, symbolized in Scilab by a backslash `\` is formally defined as:

  $$A = B \setminus C \Rightarrow B * A = C$$

- Right-matrix division (also called *feedback*), symbolized by a slash `/` in Scilab and formally defined as:

  $$A = B\,/\,C \Rightarrow B = A * C$$

Tip › *In order to prevent confusion with the matrix operation, the element-wise operation is preceded with a* `.`.

```
-->A=[2 1 1;1 2 1;1 1 2]
 A  =
    2.    1.    1.
    1.    2.    1.
    1.    1.    2.

-->x=[1;2;3]
 x  =
    1.
    2.
    3.
```

```
-->y=A*x    // matrix multiplication
 y  =
    7.
    8.
    9.

-->A\y      // =x left division
 ans  =
    1.
    2.
    3.

-->U=y/x    // right division
 U  =
    0.    0.    2.333333333
    0.    0.    2.666666667
    0.    0.    3.

-->U*x      // = y
 ans  =
    7.
    8.
    9.

-->A*A,A^2  // matrix operation
 ans  =
    6.    5.    5.
    5.    6.    5.
    5.    5.    6.
 ans  =
    6.    5.    5.
    5.    6.    5.
    5.    5.    6.

-->A.*A,A.^2  // element-wise operation
 ans  =
    4.    1.    1.
    1.    4.    1.
    1.    1.    4.
 ans  =
    4.    1.    1.
    1.    4.    1.
    1.    1.    4.

-->x*y    // error 10

    !--error 10

 Inconsistent multiplication.
```

Caution › *Do not confuse a matrix* element-wise multiplication *with a* matrix multiplication. *Multiplying matrices of incompatible sizes returns an error (10) message* Inconsistent multiplication.

Tip › *Right-matrix division can be expressed with a combination of a left matrix division and transposes:* `B/A = (A.' \ B.').'` .

If you need to perform tensor products in Scilab, use the binary operator `.*.` and the command `kron` which computes the Kronecker product of two matrices.

```
-->A=eye(2,2),B=[0 1;1 0]
 A  =
    1.    0.
    0.    1.
 B  =
    0.    1.
    1.    0.

-->A.*.A     // tensor product
 ans  =
    1.    0.    0.    0.
    0.    1.    0.    0.
    0.    0.    1.    0.
    0.    0.    0.    1.

-->kron(A,B)     // tensor product A.*.B
 ans  =
    0.    1.    0.    0.
    1.    0.    0.    0.
    0.    0.    0.    1.
    0.    0.    1.    0.
```

9.3. Element-wise and matrix functions

Scilab has all the common functions necessary to process data stored in tables:

- To find the maximum (resp. minimum) value of a matrix, use `max` (resp. `min`).
- To compute the sum (resp. product) of matrix elements, use `sum` (resp. `prod`).
- To compute the cumulative sum (resp. sumulative product) of matrix elements, use `cumsum` (resp. `cumprod`).
- To sort matrix values, use `gsort`; and to search for a value, use `find` (also see more detailed examples of use of `find` and `vectorfind` to search for a vector within a boolean matrix and Section 13.4, *Creating a sudoku*).

```
-->A=grand(1,5,'uin',0,100)
 A  =
    93.    77.    65.    93.    49.

-->M=max(A),m=min(A)     // maximum and minimum
 M  =
    93.
 m  =
    49.
```

```
-->sum(A),prod(A)    // sum and product
 ans  =
    377.
 ans  =
    2121124005.

-->cumsum(A),cumprod(A)    // cumulative sum and product
 ans  =
    93.    170.    235.    328.    377.
 ans  =
    93.    7161.    465465.    43288245.    2121124005.

-->[values,position]=gsort(A,'g','i')    // sort in ascending order
 position  =
    5.    3.    2.    1.    4.
 values  =
    49.    65.    77.    93.    93.

-->[values,position]=gsort(A,'g','d')    // sort in descending order
 position  =
    1.    4.    2.    3.    5.
 values  =
    93.    93.    77.    65.    49.

-->A(position)    // =values (sorted in order)
 ans  =
    93.    93.    77.    65.    49.

-->find(A==M)    // find the position(s) of element M in A
 ans  =
    1.    4.
```

Tip › *A lot of matrix functions take a second optional input parameter that allows the function to apply to rows or columns:*

- *With* `'r'` *(or* `1`*), the function returns a row vector with results for each column.*
- *With* `'c'` *(or* `2`*), the function returns a column vector with results for each row.*

```
-->A=grand(3,4,'uin',0,100)
 A  =
    19.    94.    89.    41.
    54.    60.    63.    44.
    17.    62.    25.    80.

-->max(A)    // maximum
 ans  =
    94.

-->max(A,'c')    // maximum per row
 ans  =
    94.
    63.
    80.
```

```
-->max(A,'r')     // maximum per column
 ans  =
    54.    94.    89.    80.
```

You can also create sets from matrices by using the function `unique`, compute the union of matrices with `union` or their intersection with `intersect`.

```
-->// creating  sets

-->E=[1 2 2 3],F=[3 4 5]
 E  =
    1.    2.    2.    3.
 F  =
    3.    4.    5.

-->E=unique(E)     // unique elements
 E  =
    1.    2.    3.

-->union(E,F)     // union
 ans  =
    1.    2.    3.    4.    5.

-->intersect(E,F)     // intersection
 ans  =
    3.
```

Elementary functions that apply to real numbers (such as `exp`, `sin`, `sqrt`, etc.), by default, operate on matrices term by term. However, for square matrices, other versions of these functions perform the matrix operation (`expm`, `sinm`, `sqrtm`, etc.). To distinguish them, an `m` is added at the end.

```
-->A=[1,2;3,4]
 A  =
    1.    2.
    3.    4.

-->B=exp(A)     // element-wise exponent operator
 B  =
    2.718281828    7.389056099
    20.08553692    54.59815003

-->log(B)
 ans  =
    1.    2.
    3.    4.

-->C=expm(A)     // matrix exponent
 C  =
    51.9689562     74.73656457
    112.1048469    164.073803
```

```
-->logm(C)
 ans  =
    1.    2.
    3.    4.

-->B=sqrt(A)     // element-wise square root
 B  =
    1.             1.414213562
    1.732050808    2.

-->B.^2     //=A
 ans  =
    1.    2.
    3.    4.

-->C=sqrtm(A)     // matrix square root
 C  =
    0.553688567 + 0.464394163i    0.806960727 - 0.212426479i
    1.210441091 - 0.318639718i    1.764129658 + 0.145754445i

-->C^2     // =A
 ans  =
    1. + 5.55112D-17i    2. - 6.93889D-17i
    3. + 2.77556D-17i    4. + 5.55112D-17i

-->sqrtm([1:4])     // error

    !--error 10000

 sqrtm: Wrong size for input argument #1: A square matrix expected.
```

Caution › *The previously mentioned matrix operations are only applicable for square matrices. For non-square matrices, they return an error (10000) which displays the message* Wrong size for input argument #1: A square matrix expected.

9.4. Solving systems of linear equations

Scilab makes it possible to readily solve systems of linear equations in different ways. To start off, set up the system of equations in matrix form $A*X=Y$. For a system with p equations and n unknows, the matrices A, X, Y have the following sizes:

$$A=\begin{pmatrix} A_{1,1} & \dots & \dots & \dots & A_{1,n} \\ \vdots & & A_{i,j} & & \vdots \\ A_{p,1} & \dots & \dots & \dots & A_{p,n} \end{pmatrix}, \quad X=\begin{pmatrix} x_1 \\ x_2 \\ \vdots \\ x_n \end{pmatrix}, \quad Y=\begin{pmatrix} y_1 \\ y_2 \\ \vdots \\ y_p \end{pmatrix}$$

In general, to solve this system, you need to use:

- the command `linsolve`

- the left division \

```
-->// system of 3 equations with 3 unknowns

-->A=[2 1 1;1 2 1;1 1 2]
 A  =
    2.    1.    1.
    1.    2.    1.
    1.    1.    2.

-->X=[1;2;3]
 X  =
    1.
    2.
    3.

-->Y=A*X
 Y  =
    7.
    8.
    9.

-->// solving for general case

-->linsolve(A,-Y)    //=X
 ans  =
    1.
    2.
    3.

-->A\Y    // left division
 ans  =
    1.
    2.
    3.

-->// system with multiple solutions

-->A=[2 1 1 1;1 2 1 1;1 1 2 1]
 A  =
    2.    1.    1.    1.
    1.    2.    1.    1.
    1.    1.    2.    1.

-->X=[1;2;3;4]
 X  =
    1.
    2.
    3.
    4.

-->Y=A*X
 Y  =
    11.
    12.
```

```
     13.

-->linsolve(A,-Y)     // solution different from X
 ans  =
     1.526315789
     2.526315789
     3.526315789
     1.894736842

-->A\Y     // different solution
 ans  =
     2.
     3.
     4.
     0.

-->// system with no solution

-->A=[2 1 1;1 2 2 ;1 1 1]
 A  =
     2.     1.     1.
     1.     2.     2.
     1.     1.     1.

-->Y=[1;2;3]
 Y  =
     1.
     2.
     3.

-->linsolve(A,-Y)     //no solution
WARNING: Conflicting linear constraints.
 ans  =
      []

-->A\Y     //  "solution" approximation using least squares approach
Warning :
matrix is close to singular or badly scaled. rcond =    0.0000D+00
computing least squares solution. (see lsq).

 ans  =
     0.181818182
     1.181818182
     0.
```

Caution › *If the system of equations has multiple solutions, Scilab returns one of them without indicating the presence of other solutions! If no solution to the system exits,* `linsolve` *returns an empty matrix* `[ ]` *and the warning message* Conflicting linear constraints! *On the other hand, a left division returns an approximate (and therefore wrong) solution which is computed using the method of least squares. In this case, Scilab displays a warning:* matrix is close to singular or badly scaled.

Tip › `linsolve(A,Y)` *solves the system* `A*X+Y=0`. *In order to solve* `A*X=Y`, *enter* `linsolve(A,-Y)`.

For square matrices, you can also multiply the first term by the inverse of matrix `A`. There are several ways to compute the inverse of `A`:

- by using `inv`
- by raising the matrix to the power of `-1`
- with right division `/`

```
-->// system of 3 equations with 3 unknowns

-->A=[2 1 1;1 2 1;1 1 2]
 A  =
    2.    1.    1.
    1.    2.    1.
    1.    1.    2.

-->X=[1;2;3]
 X  =
    1.
    2.
    3.

-->Y=A*X
 Y  =
    7.
    8.
    9.

-->// compute the inverse of A

-->I=A^(-1),inv(A),1/A
 I  =
    0.75  - 0.25  - 0.25
  - 0.25    0.75  - 0.25
  - 0.25  - 0.25    0.75
 ans  =
    0.75  - 0.25  - 0.25
  - 0.25    0.75  - 0.25
  - 0.25  - 0.25    0.75
 ans  =
    0.75  - 0.25  - 0.25
  - 0.25    0.75  - 0.25
  - 0.25  - 0.25    0.75

-->// solve the system

-->I*Y    //=X
 ans  =
    1.
    2.
    3.

-->// error 19 if there is no inverse

-->inv(ones(2,2))

    !--error 19
```

```
Problem is singular.
```

Caution › *When the matrix is square and there is no solution, the operations return the error (19) which displays the message* Problem is singular.

Tip › *In Scilab's syntax,* **1.=1** *which can cause some confusion between element-wise division and right division. Specifically,* **1./A** *does not mean the element-wise inverse of* **A** *but rather the solution to* **X*A=1**. *On the other hand,* **A.\1** *is indeed the element-wise inverse of* **A**.

```
-->A=[1 2 3];

-->// right division

-->1/A
 ans  =
    0.071428571
    0.142857143
    0.214285714

-->1./A
Warning: "1./ ..." is interpreted as "1.0/ ...". Use "1 ./ ..." for
 element wise operation
 ans  =
    0.071428571
    0.142857143
    0.214285714

-->// element-wise inverse

-->(1)./A
 ans  =
    1.     0.5     0.333333333

-->A.^(-1)
 ans  =
    1.     0.5     0.333333333

-->A.\1
 ans  =
    1.     0.5     0.333333333
```

10
Booleans

With Scilab, you can evaluate the truth of logical propositions by using boolean expressions. This attribute is very useful for programming (see Part *Programming*).

10.1. Comparison operators and logical operators

Booleans pertain to a particular Scilab type called `boolean`. They are symbolized by two constants True (`%T`) and False (`%F`), which can be combined using the usual logical operators:

- AND denoted by `&`
- OR denoted by `|`
- NOT denoted by `~`

```
-->A=%T,B=%F
 A  =
  T
 B  =
  F

-->A&B     // conjunction
 ans  =
  F

-->A|B     // disjunction
 ans  =
  T

-->~A     // negation
 ans  =
  F

-->type(A),typeof(A)     // boolean type
 ans  =
    4.
 ans  =
 boolean
```

*Tip › In boolean expressions, **%F** can be replaced with **0** and **%T** with **1** or any other nonzero integer. However, this is not recommended since it undermines the code readability. The **bool2s** command can be used to automatically replace booleans with **0**s or **1**s.*

```
-->A=~1,B=~0     // negations
 A  =
  F
 B  =
  T

-->%F|0     // =false
 ans  =
  F

-->%F&1     // =false
 ans  =
  F

-->%F|-1.1     // =true
 ans  =
  T

-->bool2s(%T)     // =1
 ans  =
    1.

-->bool2s(%F)     // =0
 ans  =
    0.
```

In practice, booleans are generated by the comparison operators:

- equal == as well as `isequal`
- inequality <> or ~=
- greater than (resp. less than) denoted by > and (resp. <)
- greater than or equal to (resp. less than or equal to) denoted by >=, (resp. <=)

You can also construct more complex boolean expressions, but take care to accurately use parentheses to prevent order of operations issues:

```
-->1==3    // this is false
 ans  =
  F

-->1<2    // this is true
 ans  =
  T

-->~((1==3)|(1<2))    // with parentheses
```

```
 ans  =
  F

-->~(1==3)|(1<2)     // without parentheses <==>(~(1==3))|(1<2)
 ans  =
  T

-->~((1==3)&(1>=2))     // with parentheses
 ans  =
  T

-->~(1==3)&(1>=2)     // without parentheses <==>(~(1==3))&(1>=2)
 ans  =
  F
```

Caution › *To verify a double inequality, use the boolean operator* **&**, *otherwise Scilab returns the error message* Undefined operation for the given operands.

```
-->x=2;

-->(1<x)&(x<3)  // correct syntax
 ans  =
  T

-->1<x<3         // wrong syntax

    !--error 144

Undefined operation for the given operands.
check or define function %b_1_s for overloading.
```

Caution › *Scilab's syntax for its operators is different from other languages. In Scilab:*

- *not equal to is not written* **!=**
- *AND is not written* **&&**
- *OR is not written* **||**

These notations in general return the error Missing operator, comma, or semicolon..

```
-->1!=2
    !--error 276
Missing operator, comma, or semicolon.
```

Tip › *Comparisons using* **%nan** *cannot be performed with* **==**. *Instead, use the command* **isnan**.

```
-->%nan==%nan     // returns %F instead of %T
 ans  =
  F

-->%nan<>%nan     // returns %T instead of %F
 ans  =
```

```
  T

-->isnan(%nan)     // accurately returns %T
 ans  =
  T

-->isnan(%inf)     // accurately returns %F
 ans  =
  F
```

*Tip › You can also compare real numbers to the values **%inf** and **-%inf**. The function **isinf** can also be used to check if a value is infinite.*

```
-->// comparing real numbers to infinity

-->%inf==%inf
 ans  =
  T

-->%inf<=%inf
 ans  =
  T

-->1<=%inf
 ans  =
  T

-->1>-%inf
 ans  =
  T

-->// using isinf

-->A=[0 %nan 1 %inf 2 -%inf %nan ]
 A  =
    0.    Nan    1.    Inf    2.   - Inf    Nan

-->A==%inf
 ans  =
  F F F T F F F

-->isinf(A)
 ans  =
  F F F T F T F
```

*A lot of commands that start with **is** let you test if a Scilab object belongs to a certain type or has a given value:*

- *for matrices, use **isempty**, **isvector***
- *for character strings, use **isalphanum**, **isascii***

***Caution** › Do not confuse the equality operator == with the assignment symbol = . For certain cases prone to confusion, Scilab replaces the assignment operator with the equality operator and returns the warning* Obsolete use of '=' instead of '=='.

```
-->1==3     // good comparison
 ans  =
  F

-->1=3     // bad comparison
Warning: obsolete use of '=' instead of '=='.
1=3     // bad comparison
   !
at line        2 of exec file called by :


 ans  =
  F
```

Caution › *Assigning values to booleans resulting from comparisons can lead to ambiguous expressions. In these cases, it is preferable to use parentheses:*

```
-->a=(1==3)     // assigning a boolean value
 a  =
  F

-->a=1==3     // ambiguous notation
 a  =
  F

-->a=1=3     // wrong notation
Warning: obsolete use of '=' instead of '=='.
a=1=3     // wrong notation
     !
at line        3 of exec file called by:


 a  =
  F
```

10.2. Boolean matrices

Matrices can be constructed from boolean entries in the same way as with real numbers. The previously mentioned logical and comparison operators are applied element by element for boolean matrices.

```
-->A=rand(2,3)
 A  =
    0.112135467    0.153121668    0.841551843
    0.685689596    0.697085060    0.406202476

-->B=rand(2,3)
 B  =
```

```
     0.409482548    0.113835969    0.561866074
     0.878412580    0.199833774    0.589617733

-->// element-wise comparison

-->C=(A<=B)
 C  =
  T F F
  T F T

-->D=(B<0.5)
 D  =
  T T F
  F T F

-->// element-wise logical operators

-->~C
 ans  =
  F T T
  F T F

-->C&D
 ans  =
  T F F
  F F F

-->C|D
 ans  =
  T T F
  T T T
```

Tip › *If you wish to compare a given matrix to the empty matrix* `[]`*, use the command* **`isempty`**.

```
-->A=rand(2,2),B=[]
 A  =
     0.504221281    0.387377877
     0.349361541    0.922289868
 B  =
     []

-->A==B
 ans  =
  F

-->isempty(A)
 ans  =
  F

-->isempty(B)
 ans  =
  T
```

Additionally, the logical operators and and or, can apply to matrices as a whole (unlike the operators & and |).

```
-->A=(0.5>rand(2,3))
 A  =
  F F T
  T T F

-->and(A)     // conjunction of values of A
 ans  =
  F

-->or(A)    // disjunction of values of A
 ans  =
  T
```

Note › *When used to compare booleans, the comparison operator == is equivalent to the logical connector* $\Leftrightarrow$.

```
-->A=[%T %T %F %F];B=[%T %F %T %F]
 B  =
  T F T F

-->A==B     // equivalence
 ans  =
  T F F T

-->(A&B)|((~A)&(~B))     // checking
 ans  =
  T F F T
```

Booleans make it possible to search within matrices. When the command find (previously mentioned in Section 9.3, *Element-wise and matrix functions*) is used on boolean matrices, it returns the indices of entries that contain %T. In general, the command find takes as arguments matrix expressions, comparison operators and logical connectors. First, the command computes the boolean matrix that results from the specified conditions and then it executes the find command which identifies the entries that contain %T.

```
-->// search

-->find([%F %F %T %F])
 ans  =
    3.

-->// vectors

-->L=grand(1,7,'uin',0,100)
 L  =
    22.    24.    61.    40.    70.    10.    8.
```

```
-->i=find(L==L(3))     //i=3
 i  =
    3.

-->i=find(L<50)     //multiple results
 i  =
    1.    2.    4.    6.    7.

-->// matrix

-->A=zeros(5,3);A(:)=15:-1:1
 A  =
    15.    10.    5.
    14.    9.     4.
    13.    8.     3.
    12.    7.     2.
    11.    6.     1.

-->// entry (1,3) = entry number 11

-->[i,j]=find(A==5)
 j  =
    3.
 i  =
    1.

-->ind=find(A==5)
 ind  =
    11.

-->// A=7,8,9

-->[i,j]=find((A-7).^2<=1)
 j  =
    2.    2.    2.
 i  =
    3.    4.    5.

-->ind=find((A-7).^2<=1)
 ind  =
    8.    9.    10.
```

Tip › *When the function* `find` *takes a matrix as input, it can output the list of entries in either one or two matrices. This is achieved by following the syntax:*

- `ind=find(A==M)` *returns a matrix* `ind` *containing all the matrix* ***entry indices*** *found.*
- `[i,j]=find(A==M)` *returns two matrices* `i` *and* `j` *which respectively contain the row number and column numbers found.*

The first syntax is better suited to vector searches.

Similarly, the command `vectorfind` looks for a row or column in a matrix.

```
-->A=[1 2 3;
-->   1 2 2;
-->   1 2 3;
-->   3 2 3]
 A  =
    1.    2.    3.
    1.    2.    2.
    1.    2.    3.
    3.    2.    3.

-->vectorfind(A,[1 2 3],'r')
 ans  =
    1.    3.

-->vectorfind(A,2*ones(4,1),'c')
 ans  =
    2.
```

Tip › *When* `find` *doesn't return any solutions, it returns an empty matrix. To check if a value* `x` *is present in a matrix* **M***, write* `find(M==x)<>[]`.

11
Character Strings and Text Files

In many situations, you will need to manipulate data that is formatted as character strings. These strings are not only used to display text. A large number of functions output results in the form of character strings (we have already dealt with some examples, for instance in chapter *Inputs/Outputs*: `pwd`, `who`, `date`, `editvar`, etc.). Other more complex functions also call for string inputs. It is therefore very important to understand character string operations not only to displays of information in the console, but also to write data to text files and create more complex programs.

11.1. Creating and displaying character strings

To get started with strings, keep in mind that:

- Character strings are delimited by single `'` or double `"` quotation marks.
- Strings can be concatenated with the symbol plus `+` (or by using `strcat` for string matrices).
- An empty string can be created by using `''` or `emptystr`.

Strings can be stored inside variables which belong to the *string* type.

```
-->A='scilab'
 A  =
 scilab

-->typeof(A)     // string type
 ans  =
 string

-->length(A)     // length of a string
 ans  =
    6.

-->B='5.4.0'
 B  =
 5.4.0

-->A+B     // concatenate
```

```
 ans  =
 scilab5.4.0

-->C=emptystr()     // empty string, equivalent to C=''
 C  =


-->length(C)     // =0
 ans  =
    0.
```

Caution › To insert a quotation mark or apostrophe within a character string, you need to precede it with an additional apostrophe, which, in this case, is used as an escape character. Otherwise, you will get an error (276) stating Missing operator, comma, or semicolon.

```
-->'Here''s an example of a string with an apostrophe'// escape
 character '
 ans  =

 Here's an example of a string with an apostrophe

-->'Here's an example of a string with an apostrophe'// error
                                                      !--error 276
Missing operator, comma, or semicolon.
```

As with integers and booleans, you can create matrices of strings by using the usual matrix operations. When these matrices are displayed, they are delimited on the left and right by exclamation points `!`, to distinguish them from matrices containing other types of objects. The command `length` computes the length of a string.

```
-->A='scilab'
 A  =
 scilab

-->B='5.5.2'
 B  =
 5.5.2

-->M=[A, B; B,A]        // string matrix
 M  =
!scilab  5.5.2   !
!                !
!5.5.2   scilab  !

-->length(M)          // length of strings in M
 ans  =
    6.    5.
    5.    6.
```

Caution › When used on a string matrix, the `length` *command doesn't return the length of the matrix but rather outputs a matrix containing the length of each string contained within the matrix!*

All object types available in Scilab can be converted to a string with the `string` command, and even converted to LaTeX or MathML with the command `prettyprint`. To display these strings in the console, use the commands `disp` or `printf`.

```
-->e=string(%e)     // convert to a string
 e  =
 2.718281828

-->// display real numbers

-->disp("pi="+string(%pi)+"..., e="+string(e)+'...')

 pi=3.141592654..., e=2.718281828...

-->printf("pi=%f..., e=%f...\n",%pi,%e)
pi=3.141593..., e=2.718282...

-->// display integers

-->printf(" un=%d \n deux=%d \n trois=%d\n",1,2,3)
 un=1
 deux=2
 trois=3

-->prettyprint(eye(2,2))    // identity matrix in LaTeX
 ans  =
 ${\begin{pmatrix}1&0\cr 0&1\cr \end{pmatrix}}$
```

Tip › `printf` *allows the same syntax as the C function of the same name. In a string,*

- `\n` *produces a new line;*
- `\r` *indicates a carriage return (without producing a new line);*
- `\t` *indicates a tab;*
- `%s` *gets replaced by a string;*
- `%d` *gets replaced by an integer;*
- `%f` *gets replaced by a double.*

Type `help printf_conversion` *for more information on display formats for different types of variables.*

To resolve issues related to text justification in strings, use the function `justify` which lets you center, left-align or right-align the text.

```
-->M=['short text';'     text with spaces        '; 'loooooonnnnnngggg
 text']
 M  =
!short text                    !
!                              !
!     text with spaces         !
!                              !
```

```
!loooooonnnnnnngggg text       !

-->justify(M,'l')  // left-alignment
 ans  =
!short text                    !
!                              !
!text with spaces              !
!                              !
!loooooonnnnnnngggg text       !

-->justify(M,'r')  // right-alignment
 ans  =
!               short text     !
!                              !
!         text with spaces     !
!                              !
!loooooonnnnnnngggg text       !

-->justify(M,'c')  // center
 ans  =
!       short text             !
!                              !
!    text with spaces          !
!                              !
!loooooonnnnnnngggg text       !
```

11.2. Manipulating strings

Scilab provides a large number of functions that are very useful to manipulate character strings and often mimic functions from other languages:

- To relate a character to its position in a string, use `strindex` and `part`.
- To substitute characters from strings, use `strsubst`.
- To split strings into substrings, use `tokens` or `strsplit`.
- To search for substrings, use `grep` or `regex`.

Go to the appropriate help chapter by typing `apropos 'Strings'` to get a complete list of these functions.

```
-->str='This is a character string'
 str  =
 This is a character string

-->n=strindex(str,'i')    // position of character 'i' in str
 n  =
    3.    6.    24.

-->part(str,n)    // string in position n
 ans  =
 iii
```

```
-->strsubst(str,'i','*')    // replace 'i' with '*'
 ans  =
 Th*s *s a character str*ng

-->tokens(str)    // split according to the delimiter ' '
 ans  =
!This       !
!           !
!is         !
!           !
!a          !
!           !
!character  !
!           !
!string     !

-->tokens(str,'i')    // split according to the delimiter  'i'
 ans  =
!Th                 !
!                   !
!s                  !
!                   !
!s a character str  !
!                   !
!ng                 !

-->str='scilab'
 str  =
 scilab

-->strsplit(str)    // split characters
 ans  =
!s  !
!   !
!c  !
!   !
!i  !
!   !
!l  !
!   !
!a  !
!   !
!b  !

-->strsplit(str,'i')    // split according to delimiter 'i'
 ans  =
!sc   !
!     !
!lab  !

-->str = ["hat" "cat" "chat" "tac" "dog"]
 str  =
!hat  cat  chat  tac  dog  !

-->grep(str,'a')
```

```
 ans  =
    1.    2.    3.    4.

-->str='aababbbaaabba'
 str  =
 aababbbaaabba

-->[first,last,match]=regexp(str,'/a(b)+/')
 match  =
!ab    !
!      !
!abbb  !
!      !
!abb   !
 last  =
    3.    7.    12.
 first  =
    2.    4.    10.
```

You can convert character strings to vectors of ASCII codes and vice versa with the command `ascii`.

```
-->tab=ascii('Scilab-5.5.0')
 tab  =
    83.    99.    105.    108.    97.    98.    45.    53.    46.    53.
    46.    48.

-->str=ascii(tab)
 str  =
 Scilab-5.5.0
```

It is possible to evaluate strings that correspond to real numbers or any other Scilab variable type by using the function `evstr`. Likewise, a Scilab command stored within a string can be executed with the command `execstr`.

```
-->a=evstr('123/2')    // evaluate a string
 a  =
    61.5

-->execstr('b=2*11')    // execute a string

-->b
 b  =
    22.
```

Tip › It is often very convenient to generate strings containing lengthy instructions to later evaluate them by using **execstr** *:*

```
-->num=string([1 2 3]')
 num  =
!1  !
!   !
!2  !
!   !
!3  !

-->prog='x'+num+'=2*'+num // string describing a set of instructions
 prog  =
!x1=2*1  !
!        !
!x2=2*2  !
!        !
!x3=2*3  !

-->execstr(prog)     // execute the instructions contained in "prog"

-->x2     // variable x2 was created and has the value 4
 ans  =
    4.
```

Finally, strings are very useful to read or write to text files. The most simple commands are:

- reading a text file with the command `read` or `mgetl`
- writing to a text file with the command `write` or `mputl`

```
-->str=['scilab';strcat(string(1:10),';')]
 str  =
!scilab                !
!                      !
!1;2;3;4;5;6;7;8;9;10  !

-->mputl(str,'myfile.txt')
 ans  =
  T

-->str2=mgetl('myfile.txt')
 str2  =
!scilab                !
!                      !
!1;2;3;4;5;6;7;8;9;10  !
```

Other functions exist, such as `mfprintf`, `mopen`, `mclose`, which copy the procedure of the C language functions of the same name.

***Caution* › `read` and `write` emulate the FORTRAN commands and make it possible to read and write Scilab variables in a text file. However, `write` cannot rewrite to an existing file.**

```
-->unix('DEL myfile.txt');

-->x=[1:4]';y=cos(x);

-->write('myfile.txt',[x y])

-->str2=read('myfile.txt',-1,2)
 str2  =
    1.      0.540302306
    2.    - 0.416146837
    3.    - 0.989992497
    4.    - 0.653643621
```

The commands `read` and `write` can also be used to read and write to the standard input and output. In Scilab, the standard input and output are stored inside the variable `%io` and indexed by two integers.

```
-->%io
 %io  =
    5.    6.

-->// write to the standard output

-->write(%io(2),'hello')
hello

-->// read a string of size 1x1 from the standard input

-->txt=read(%io(1),1,1,'(a)')
 txt  =
 hello

-->typeof(txt)
 ans  =
 string
```

12

Other Common Types

Other data types can be manipulated in Scilab, for example polynomials, rational functions, lists and hypermatrices. You can also create your own data type, as we will see in Part *Programming*.

12.1. Polynomials

Polynomials in a single indeterminate define the Scilab *polynomial* type. The command `poly` can be used to create a polynomial by defining its roots or its coefficients.

*Tip › `poly([a b], 'x')` returns the polynomial `(x-a)(x-b)`, that is, the smallest monic polynomial with roots **a** and **b**.*

Polynomials can be combined by using the elementary operations `+`, `-`, `*`, `^`, `/`. The following more complex operations can also be performed:

- retrieve a vector containing the polynomial coefficients with `coeff`
- compute the degree of a polynomial with `degree`
- evaluate a polynomial at one or several points with `horner`
- compute the roots (real as well as complex) of a polynomial with `roots`
- factor polynomials with `factors`
- compute the GCD or the LCM of two polynomials with `gcd` and `lcm`.

Caution *› Scilab is not a system that performs symbolic computation. The polynomial coefficients are double precision real numbers. Due to rounding errors, computations can return incorrect results!*

```
-->X=poly(0,'x')    // variable 'x'
 X  =
    x

-->P=X^2-3*X+2      // combination of polynomials
 P  =
              2
    2 - 3x + x

-->coeff(P)         // coefficients
 ans  =
    2.  - 3.    1.
```

```
-->degree(P)        // degree =2
 ans  =
    2.

-->horner(P,[0:3]) // evaluating
 ans  =
    2.    0.    0.    2.

-->roots(P)         // roots
 ans  =
    2.
    1.

-->Q=poly([-4 0 1],'x','coeff')
 Q  =
          2
  - 4 + x

-->gcd([P,Q])             // =X-2
 ans  =
  - 2 + x

-->lcm([P,Q])             // =(x^2-4)*(x-1)
 ans  =
              2   3
    4 - 4x - x + x

-->[pgcd,U]=bezout(P,Q)  // Bézout's identity
 U  =
  - 0.333333333     2 + x

    0.333333333     1 - x
 pgcd  =
  - 2 + x

-->L=factors(P),          // P(x)=(x-1)*(x-2)
 L  =

       L(1)

  - 2 + x

       L(2)

  - 1 + x
```

Tip › The function **horner** *can also compute the composition of polynomials.*

```
-->P=poly(1,'x')      // P(x)=x-1
 P  =
  - 1 + x

-->Q=poly([-1 1],'x')     // Q(x)=x^2-1
 Q  =
```

```
            2
  - 1 + x

-->horner(P,Q),Q-1      // P(Q(x))=x^2-1
 ans  =
            2
  - 2 + x
 ans  =
            2
  - 2 + x

-->horner(Q,P),P^2-1      // Q(P(x))=x^2-2*x
 ans  =
             2
  - 2x + x
 ans  =
             2
  - 2x + x
```

Caution › *Scilab cannot manipulate polynomials in multiple indeterminates. If you combine polynomials with different indeterminates, you will get an error (144) stating* Undefined operation for the given operands.

```
-->X=poly(0,'x')
 X  =
    x

-->Z=poly(0,'z')
 Z  =
    z

-->X+Z     // error 144

    !--error 144

Undefined operation for the given operands.
check or define function %p_a_p for overloading.
```

12.2. Rational fractions

Rational fractions can be constructed from polynomials and define the type *rational*. The commands `denom` and `numer` return the denominator and numerator. As with other data types, you can construct matrices of rational fractions (or polynomials). For these matrices, an additional specific command is available to compute the inverse: `invr`.

```
-->P=poly(1,'x')      // variable 'x'
 P  =
  - 1 + x

-->Q=poly(2,'x')      // combination of polynomials
 Q  =
```

```
   - 2 + x

-->R=P/Q    // fraction  (x-1)/(x-2)
 R  =
  - 1 + x
    -----
  - 2 + x

-->numer(R)    // numerator
 ans  =
  - 1 + x

-->denom(R)    // denominator
 ans  =
  - 2 + x

-->M=[1 R; P Q]    // matrix of rational fractions
 M  =
    1       - 1 + x
    -         -----
    1       - 2 + x

  - 1 + x   - 2 + x
    -----     -----
      1         1

-->N=invr(M)    // inverse matrix
 N  =
                2
    4 - 4x + x      1 - x
    ----------      -----
      3 - 2x        3 - 2x

                2
  - 2 + 3x - x    - 2 + x
    ----------      -----
      3 - 2x        3 - 2x

-->N*M    // =identity matrix
 ans  =
    1     0
    -     -
    1     1

    0     1
    -     -
    1     1

-->// computing the characteristic polynomial

-->X=poly(0,'x');

-->A=[1 2; 3 4]
 A  =
    1.    2.
    3.    4.
```

```
-->P=det(A-X*eye(2,2))
 P  =
                2
  - 2 - 5x + x
```

12.3. Lists

You will also very often come across the *list* type in Scilab.

Creating and manipulating lists

The basic commands include :

- creating a list of different objects with `list`
- creating a list by concatenating different lists with `lstcat`
- deleting an element in a list with `null`

Lists are then manipulated the same way as vectors. In particular, the command `length` computes the length of a list (the number of elements).

```
-->list() // empty list
 ans  =
     ()

-->// list with 4 elements:

-->L=list(%pi, eye(2,2), poly(0,'x'), 'scilab')
 L  =

       L(1)

    3.141592654

       L(2)

    1.    0.
    0.    1.

       L(3)

    x

       L(4)

 scilab

-->typeof(L) // list type
 ans  =
```

```
 list

-->L(1) // retrieve
 ans  =
    3.141592654

-->L(1)=null() // delete
 L  =

       L(1)

    1.    0.
    0.    1.

       L(2)

    x

       L(3)

 scilab

-->length(L)
 ans  =
    3.

-->M=list(%e,rand(1,2),'end')
 M  =

       M(1)

    2.718281828

       M(2)

    0.685397966    0.890622473

       M(3)

 end

-->N=lstcat(L,M) // concatenating lists
 N  =

       N(1)

    1.    0.
    0.    1.

       N(2)

    x

       N(3)

 scilab
```

```
        N(4)

     2.718281828

        N(5)

     0.685397966    0.890622473

        N(6)

  end
```

Caution › *The command* `L=null()` *does not delete the first element in the list but removes the variable L! This method can be used to delete any type of variable similarly to the command* `clear`.

```
-->L=list(1)    // list with one element
 L  =

        L(1)

     1.

-->L(1)=null() // empty list
 L  =
     ()

-->L=list(1,2) // list with two elements
 L  =

        L(1)

     1.

        L(2)

     2.

-->L=null()     // deletes the variable L

-->L            // returns an error

     !--error 4

 Undefined variable: L
```

Displaying lists as arrays

Lists can be displayed as tables by using the command `cell`. By default, this command creates an empty array of lists which can then be filled with objects of differents types:

```
-->L=cell()  // empty array of lists
 L  =

     []

-->L=cell(2,3)  // array of lists of size 2x3
 L  =

!{}  {}  {}  !
!            !
!{}  {}  {}  !

-->// filling L with objects of different types

-->L(1,1).entries="Scilab"
 L  =

!"Scilab"  {}  {}  !
!                  !
!{}        {}  {}  !

-->L(1,2).entries=%i
 L  =

!"Scilab"  %i  {}  !
!                  !
!{}        {}  {}  !

-->L(1,3).entries=%T
 L  =

!"Scilab"  %i  %t  !
!                  !
!{}        {}  {}  !

-->L(2,1).entries=eye(2,2)
 L  =

!"Scilab"        %i  %t  !
!                        !
!{2x2 constant}  {}  {}  !

-->L(2,2).entries=poly(1,'x')
 L  =

!"Scilab"        %i    %t  !
!                          !
!{2x2 constant}  -1+x  {}  !

-->L.dims,size(L)  // size of L
 ans  =
  2  3
 ans  =
    2.    3.
```

```
-->typeof(L),type(L) // type 17 =ce
 ans  =
 ce
 ans  =
    17.

-->L(2,1),typeof(L(2,1)) // L(2,1) of type ce
 ans  =

 {2x2 constant}
 ans  =
 ce

-->// L(2,1).entries of type constant :

-->L(2,1).entries,typeof(L(2,1).entries)
 ans  =
    1.    0.
    0.    1.
 ans  =
 constant
```

Manipulating a `cell` is very similar to dealing with matrices. The `cells` belong to the data type `ce` (17).

***Caution** › Each entity in a **`cell`** is a cell in itself! To retrieve an object stored inside the sub-cell **`L(i,j)`** of a cell, enter the command **`L(i,j).entries`**. This can be verified with the command **`iscell`** which returns the boolean **%T** for objects belonging to the cell type and otherwise returns **%F**.*

```
-->L=cell(2,2) // list array of size 2x2
 L  =

!{}  {}  !
!        !
!{}  {}  !

-->L(1,1).entries="Scilab"
 L  =

!"Scilab"  {}  !
!              !
!{}        {}  !

-->iscell(L)
 ans  =
  T

-->iscell(L(1,1))
 ans  =
  T

-->iscell(L(1,1).entries)
 ans  =
  F
```

Indexing fields with character strings

Entries of a list can be indexed with character strings rather than numerical indices by using the command `struct`. These are called *structures* which are also found in the C language. Structures belong to the type 17 and are also denoted by `st`.

```
-->endoftheworld=struct('day',21,'month',12,'year',2012)  // creating
 endoftheworld  =
   day: 21
   month: 12
   year: 2012

-->typeof(endoftheworld),type(endoftheworld)  // type st = type 17
 ans  =
 st
 ans  =
    17.

-->endoftheworld.day  // retrieve
 ans  =
    21.

-->endoftheworld('month')
 ans  =
    12.

-->endoftheworld.year=2013  // modify
 endoftheworld  =
   day: 21
   month: 12
   year: 2013

-->isstruct(endoftheworld) // true
 ans  =
  T

-->isstruct(1)  // false
 ans  =
  F
```

Typed lists

Finally, there also exists a type for particular lists called *typed lists*, which can be used to create new types in Scilab. They can be created with the commands `tlist` and `mlist`.

12.4. Hypermatrices

Matrices of larger than two dimensions can also be manipulated in Scilab and are called *hypermatrices*. To create hypermatrices, use the command `hypermat`. Filling the matrices can be done during their creation or later by assigning the different sub-matrices. In general, hypermatrices are manipulated the same way as matrices. For example, the command `size` returns the size of the hypermatrix (as a vecteur) and `ndims` returns the number of dimensions of the hypermatrix (i.e. the length the output of `size`).

```
-->M =  hypermat([2 3 4])    // 4 matrices of 2 rows and 3 columns
 M  =
(:,:,1)

    0.    0.    0.
    0.    0.    0.
(:,:,2)

    0.    0.    0.
    0.    0.    0.
(:,:,3)

    0.    0.    0.
    0.    0.    0.
(:,:,4)

    0.    0.    0.
    0.    0.    0.

-->M=M+1     // scalar addition
 M  =
(:,:,1)

    1.    1.    1.
    1.    1.    1.
(:,:,2)

    1.    1.    1.
    1.    1.    1.
(:,:,3)

    1.    1.    1.
    1.    1.    1.
(:,:,4)

    1.    1.    1.
    1.    1.    1.

-->ndims(M)     // hypermatrix with 3 dimensions
 ans  =
    3.

-->size(M)     // size 2x3x4
```

```
 ans  =
    2.    3.    4.

-->typeof(M)    // type hypermat
 ans  =
 hypermat

-->N=  hypermat([2 3 4],1:24)
 N  =
(:,:,1)

    1.    3.    5.
    2.    4.    6.
(:,:,2)

    7.    9.     11.
    8.    10.    12.
(:,:,3)

    13.    15.    17.
    14.    16.    18.
(:,:,4)

    19.    21.    23.
    20.    22.    24.

-->N(:,:,3)     // 3rd "sub-matrix"
 ans  =
    13.    15.    17.
    14.    16.    18.

-->M+N     // addition
 ans  =
(:,:,1)

    2.    4.    6.
    3.    5.    7.
(:,:,2)

    8.    10.    12.
    9.    11.    13.
(:,:,3)

    14.    16.    18.
    15.    17.    19.
(:,:,4)

    20.    22.    24.
    21.    23.    25.
```

Note that you can generate hypermatrices in the same way as matrices, with the functions `zeros`, `ones` and `rand`.

```
-->rand(2,2,2)
 ans  =
(:,:,1)

    0.948818426    0.376011873
    0.343533725    0.734094056
(:,:,2)

    0.261576147    0.263857842
    0.499349384    0.525356309

-->ones(2,2,2)
 ans  =
(:,:,1)

    1.    1.
    1.    1.
(:,:,2)

    1.    1.
    1.    1.

-->zeros(2,2,2)  // equivalent to hypermat([2 2 2])
 ans  =
(:,:,1)

    0.    0.
    0.    0.
(:,:,2)

    0.    0.
    0.    0.
```

Tip* › *In the RGB system, color images are defined by three matrices representing the intensity levels of the three primary colors for each pixel. Each image is thus coded as a hypermatrix of size* `3 x rows x columns`*. This is performed with the SIVP module.

13 Calculation Examples

The matrix format used by Scilab makes it possible to perform complicated calculations in just a few lines and with a reduced computational time compared to more classical programming using loops and conditional structures (see Section 17.2, *Function optimization* in the Part *Programming*). Here are a few examples to familiarize yourself with these techniques.

13.1. Creating vectors and matrices

Our aim is to construct them by using the least amount of commands possible in Scilab.

List multiples of 3 between 1 and 21

Start from 3, increment by 3:

```
-->u1=[3:3:21]
 u1  =
    3.    6.    9.    12.    15.    18.    21.
```

List odd integers between 10 and 20

Start from 11, increment by 2:

```
-->u2=[11:2:20]
 u2  =
    11.    13.    15.    17.    19.
```

List the squares of integers from −5 to 3

Square the list of integers from −5 to 3:

```
-->u3=[-5:3].^2
 u3  =
    25.    16.    9.    4.    1.    0.    1.    4.    9.
```

List the squares of integers from 3 to −5

Reverse the order of the list u3 or follow the same method starting from a vector of integers from 5 to −3:

```
-->u4=[3:-1:-5].^2          //  first solution
 u4  =
    9.    4.    1.    0.    1.    4.    9.    16.    25.

-->u4=u3($:-1:1)            //  second solution
 u4  =
    9.    4.    1.    0.    1.    4.    9.    16.    25.
```

List numbers of the form $2n+3$ for n between 2 and 10

Create a list of integers n from 2 to 10, then use the element-wise operations + and *, or use the same method as for u2:

```
-->n=[2:10];u5=2*n+3  // first solution
 u5  =
    7.    9.    11.    13.    15.    17.    19.    21.    23.

-->u5=2*[2:10]+3      // using one command
 u5  =
    7.    9.    11.    13.    15.    17.    19.    21.    23.

-->u5=[7:2:23]        // second solution
 u5  =
    7.    9.    11.    13.    15.    17.    19.    21.    23.
```

List the multiplicative inverses of integers from 2 to 9

Create a list of integers from 2 to 9, then use the element-wise division ./:

```
-->u6=(1)./[2:9]   //  warning:  1./ = (1.)/ =/= (1)./
 u6  =
    0.5    0.333333333    0.25    0.2    0.166666667    0.142857143
   0.125    0.111111111

-->u6=[2:9].^(-1)  //  second solution
 u6  =
    0.5    0.333333333    0.25    0.2    0.166666667    0.142857143
   0.125    0.111111111
```

List the powers of 2 for an exponent from 2 to 9

Use the same method as for u6 with the power sign ^:

```
-->u7=2^[2:9]
 u7  =
    4.    8.    16.    32.    64.    128.    256.    512.
```

List the values of $\frac{2^n}{n}$ for n from 2 to 9

Perform an element-wise division of the list u7 with the list of integers from 2 to 9:

```
-->u8=u7./[2:9]          // first solution
 u8  =
    2.    2.666666667    4.    6.4    10.66666667    18.28571429
 32.    56.88888889

-->n=[2:9];u8=(2^n)./n  // second solution
 u8  =
    2.    2.666666667    4.    6.4    10.66666667    18.28571429
 32.    56.88888889
```

List of special angles $u9 = 0 \quad \pi/6 \quad \pi/4 \quad \pi/3 \quad \pi/2 \quad \pi$

We start by constructing a list of inverses of integers from 6 to 1, then deal with each particular case (adding the angle 0 and deleting $\frac{\pi}{5}$):

```
-->u9=%pi*[0 1/6 1/4 1/3 1/2 1]           //  first solution
 u9  =
    0.    0.523598776    0.785398163    1.047197551    1.570796327
  3.141592654

-->u9=(%pi)./[6:-1:1],u9(2)=[],u9=[0 u9]   //  second solution
 u9  =
    0.523598776    0.628318531    0.785398163    1.047197551
 1.570796327    3.141592654
 u9  =
    0.523598776    0.785398163    1.047197551    1.570796327
 3.141592654
 u9  =
    0.    0.523598776    0.785398163    1.047197551    1.570796327
  3.141592654
```

List of sine values of special angles

Apply the function sin to the previous list:

```
-->u10=sin(u9)
 u10  =
    0.    0.5    0.707106781    0.866025404    1.    1.22465D-16
```

List containing the number 0 six times

You can use the function zeros, or use a combination of vectors:

```
-->u11=zeros(1,6)    // first solution
 u11  =
    0.    0.    0.    0.    0.    0.

-->u11=0*[1:6]       // second solution
 u11  =
    0.    0.    0.    0.    0.    0.

-->u11=[1:6]-[1:6]   // third solution
 u11  =
    0.    0.    0.    0.    0.    0.
```

List containing 6 times the number 1

Use the function ones or a combination of vectors:

```
-->u12=ones(1,6)     // first solution
 u12  =
    1.    1.    1.    1.    1.    1.

-->u12=1+zeros(1,6) // second solution
 u12  =
    1.    1.    1.    1.    1.    1.

-->u12=[1:6]-[0:5]   // third solution
 u12  =
    1.    1.    1.    1.    1.    1.
```

List 6 times the number 0, then 6 times the number 1

Concatenate u12 and u11:

```
-->u13=[u11 u12]
 u13  =
    0.     0.     0.     0.     0.     0.     1.     1.     1.     1.     1.
  1.
```

Matrix containing 6 times the number 1 on the first row and 6 times the number 0 below

Here, also concatenate u12 and u11:

```
-->u14=[u12;u11]
 u14  =
    1.     1.     1.     1.     1.     1.
    0.     0.     0.     0.     0.     0.
```

13.2. Solving calculations related to series

In mathematics, a *series* is a sum of the form $\sum_{k=1}^{\infty} f(x)$ and its partial sum of order $n \in \mathrm{N}^*$ is $\sum_{k=1}^{n} f(x)$. Sequences and series provide interesting examples of calculations and searches in matrices. Here, we are going to perform computations related to the series $\sum_{k=1}^{\infty} \frac{1}{k^2} = \frac{\pi^2}{6}$ with a minimum possible number of commands.

Compute the partial sum up to the term $n = 10^5$

Start by creating a list of multiplicative inverses of the squares of integers from 1 to to *n*, then apply the sum command to the matrix obtained:

```
-->n=1D5;%pi^2/6              // exact value
 ans  =
    1.644934067

-->F1=[1:n].^(-2);              // list of 1/k^2

-->S1=sum(F1)                 // value of the series
 S1  =
    1.644924067
```

Compute the term for which we obtain a precision of $\varepsilon = 10^{-4}$

Assume that $n_0 \leq 10^5$ and compute all the partial sums from 1 to 1D+5 with `cumsum`. Then use `find` to search for the value of *n* which satisfies the condition:

```
-->eps=1D-4;                                // precision

-->L1=cumsum(F1);                           // list of partial sums

-->n=find(abs(L1-%pi^2/6)<=eps,1)           // term that achieves a precision
eps
 n  =
    10000.

-->L1(n),%pi^2/6                            // verifying
 ans  =
    1.644834072
 ans  =
    1.644934067
```

13.3. Creating a complicated matrix

Matrix multiplication can also be used to create complicated matrices in just a few commands. One particular type of matrix multiplication only involves row/column vectors:

$$\begin{pmatrix} b_1 & \cdots & b_j & \cdots & b_n \end{pmatrix} = B$$

$$A = \begin{pmatrix} a_1 \\ \vdots \\ a_i \\ \vdots \\ a_p \end{pmatrix} \quad \begin{pmatrix} a_1b_1 & \cdots & a_1b_j & \cdots & a_1b_n \\ \vdots & \vdots & \vdots & \vdots & \vdots \\ a_ib_1 & \cdots & a_ib_j & \cdots & a_ib_n \\ \vdots & \vdots & \vdots & \vdots & \vdots \\ a_pb_1 & \cdots & a_pb_j & \cdots & a_pb_n \end{pmatrix} = A * B$$

By performing a matrix multiplication of vectors, we obtain the following matrices of size `n=6` rows by `m=5` columns:

`L(i,j)=i=` row number of element `(i,j)`

Start by constructing

1. one column vector `A` with 6 lines such that `A(i)=i` by using the `:` and `'` operators ;
2. one row vector `B` with 5 columns such that `B(j)=1` by using the command `ones` ;

then multiply both vectors in the correct order by following the matrix multiplication formula `L(i,j)=A(i)*B(j)=i`. We obtain:

```
-->n=6;m=5;

-->A=[1:n]'
 A  =
    1.
    2.
    3.
    4.
    5.
    6.

-->B=ones(1,m)
 B  =
    1.     1.     1.     1.     1.

-->L=A*B
 L  =
    1.     1.     1.     1.     1.
    2.     2.     2.     2.     2.
    3.     3.     3.     3.     3.
    4.     4.     4.     4.     4.
    5.     5.     5.     5.     5.
    6.     6.     6.     6.     6.

-->// with only one command

-->L=([1:n]')*ones(1,m)
 L  =
    1.     1.     1.     1.     1.
    2.     2.     2.     2.     2.
    3.     3.     3.     3.     3.
    4.     4.     4.     4.     4.
    5.     5.     5.     5.     5.
    6.     6.     6.     6.     6.
```

`C(i,j)=j`= column number of element `(i,j)`

Use the previous method to compute the product `C(i,j)=A(i)*B(j)=j`:

```
-->n=6;m=5;

-->K=ones(n,1)*[1:m]     // column number
 K  =
    1.     2.     3.     4.     5.
    1.     2.     3.     4.     5.
    1.     2.     3.     4.     5.
    1.     2.     3.     4.     5.
    1.     2.     3.     4.     5.
    1.     2.     3.     4.     5.
```

`D(i,j)=1+|i-j|`= diagonal index of element `(i,j)`

A matrix `D` is defined by a formula which depends on its number of rows and columns `i` and `j`. Therefore, we can easily use matrix operations or element-wise operations to combine the matrices `L` and `C` obtained in the previous example.

```
-->n=6;m=5;

-->L=([1:n]')*ones(1,m); // row number

-->C=ones(n,1)*[1:m];    // column number

-->D=1+abs(L-C)          // diagonal index
 D  =
    1.    2.    3.    4.    5.
    2.    1.    2.    3.    4.
    3.    2.    1.    2.    3.
    4.    3.    2.    1.    2.
    5.    4.    3.    2.    1.
    6.    5.    4.    3.    2.
```

`N(i,j)=i+n*(j-1)` position number of element `(i,j)`

(position related to the concatenation of columns)

Here again, the simplest solution is to use the matrices `L` et `C`, but you can also take advantage of certain attributes of the `:` operator.

```
-->n=6;m=5;

-->L=([1:n]')*ones(1,m); // line number

-->C=ones(n,1)*[1:m];    // column number

-->// N = matrix of element positions

-->N=L+(C-1)*n          // method 1
 N  =
    1.    7.     13.    19.    25.
    2.    8.     14.    20.    26.
    3.    9.     15.    21.    27.
    4.    10.    16.    22.    28.
    5.    11.    17.    23.    29.
    6.    12.    18.    24.    30.

-->N=zeros(n,m);N(:)=[1:n*m] // method 2
 N  =
    1.    7.     13.    19.    25.
    2.    8.     14.    20.    26.
```

```
     3.     9.     15.     21.     27.
     4.     10.    16.     22.     28.
     5.     11.    17.     23.     29.
     6.     12.    18.     24.     30.
```

13.4. Creating a sudoku

A sudoku is a grid of size 9×9 whose squares are filled with numbers from 1 to 9 with each integer appearing only once in each row, column and *region*. In practice, only part of the grid's content is revealed and the purpose of the puzzle is to uncover the values of the remaining squares:

$$S = \begin{array}{|c|c|c||c|c|c||c|c|c|}\hline 5 & 3 & & & 8 & & & 2 & \\ \hline 8 & & & & 4 & 2 & & & \\ \hline & & 1 & 3 & & 6 & & 8 & \\ \hline\hline 6 & 5 & 3 & & & & 1 & & 2 \\ \hline 2 & 1 & 4 & 6 & & 3 & 5 & 7 & 8 \\ \hline 9 & & 8 & & & & 3 & 6 & 4 \\ \hline\hline & 6 & & 5 & & 1 & 8 & & \\ \hline & & & 4 & 6 & & & & 5 \\ \hline & 4 & & & 3 & & & 1 & 6 \\ \hline \end{array}$$

Representation of a sudoku using a matrix in Scilab

You can fill the empty entries with 0s:

```
-->     8  0  0  0  4  2  0  0  0;
-->     0  0  1  3  0  6  0  8  0;
-->     6  5  3  0  0  0  1  0  2;
-->     2  1  4  6  0  3  5  7  8;
-->     9  0  8  0  0  0  3  6  4;
-->     0  6  0  5  0  1  8  0  0;
-->     0  0  0  4  6  0  0  0  5;
-->     0  4  0  0  3  0  0  1  6];
```

Defining the region associated with an element

Create a matrix `R` and assign a number to each region by using the formula `R(i,j)=1+int((j-1)/3)+int((i-1)/3)*3` where `i` and `j` are the row/column indices of each element:

```
-->n=9;m=9;                              // size of matrices

-->L=([1:n]')*ones(1,m);                 // row number

-->K=ones(n,1)*[1:m];                    // column number

-->R=1+int((K-1)/3)+int((L-1)/3)*3 // region number
 R  =
    1.    1.    1.    2.    2.    2.    3.    3.    3.
    1.    1.    1.    2.    2.    2.    3.    3.    3.
    1.    1.    1.    2.    2.    2.    3.    3.    3.
    4.    4.    4.    5.    5.    5.    6.    6.    6.
    4.    4.    4.    5.    5.    5.    6.    6.    6.
    4.    4.    4.    5.    5.    5.    6.    6.    6.
    7.    7.    7.    8.    8.    8.    9.    9.    9.
    7.    7.    7.    8.    8.    8.    9.    9.    9.
    7.    7.    7.    8.    8.    8.    9.    9.    9.
```

Note that here, once more thanks to matrix multiplication, there was no need to use loops.

Retrieving values already used in a row/column/region

Use `find` to locate the elements of matrix `S` that do not contain a 0. You can restrict your search to a given row/column/region by using the `L`, `C` and `R` matrices we previously created. To avoid locating the same value multiple times, use the `unique` command on the results:

```
-->i=3;j=5;  // element of interest

-->// values already using in the row

-->A=unique(S(find((S<>0)&(L==i))))
 A  =
    1.
    3.
    6.
    8.

-->// values already used in the column

-->B=unique(S(find((S<>0)&(K==j))))
 B  =
    3.
    4.
    6.
    8.

-->// values already used in the region

-->C=unique(S(find((S<>0)&(R==R(i,j)))))
```

```
 C  =
    2.
    3.
    4.
    6.
    8.
```

The result is directly transfered to `S` to retrieve the values already used.

Retrieving possible values for an element `(i,j)=(3,5)`

In order to get a list of possible values for element `(i,j)`, create a set of all values found for the row, column and region of element`(i,j)`. Then find the complement of this set by deleting the corresponding entries of the matrix `[1:9]`:

```
-->// set of values already taken

-->D=union(union(A,B),C)
 D  =
    1.    2.    3.    4.    6.    8.

-->// candidate values for element (i,j)

-->E=[1:9];E(D)=[]
 E  =
    5.    7.    9.
```

Programming

To take full advantage of Scilab's capabilities and go beyond the simple calculations shown in Part *Computing*, you must turn to programming. Scilab has a simple programming language, which is nevertheless comprehensive enough to create complex programs. This is what we will study in this part.

14 Scripts

When you use a numerical computational software, you often are faced with the need to repeat the same calculation several times while modifying only some of the parameters. Therefore, it is convenient to be able to save steps of a complicated calculation to later come back and modify specific commands. For this purpose, Scilab provides the ability to write scripts which can be easily executed and modified from the text editor, as seen in chapter *Preview of Scilab* with the pendulum simulation or planetary motion. Here, we define a script as a list of instructions stored within a text file. These instructions can then be executed sequentially using Scilab.

14.1. Writing and executing scripts

In order to understand what a Scilab script is, we are going to look at a short example:

```
1. A=[1 2;3 4];y=[3;5];
2. x1=linsolve(A,-y);      // not displayed, even "with echo"
3. x2=A^(-1)*y             // displayed if using "with echo"
4. disp(x1,'x=')           // displayed, even if using "with no echo"
```

This script is four lines long and contains the following instructions:

1. Two instructions which define the matrices `A` and `y`.
2. Computes the solution to the system of equations $A*x=y$ by using the `linsolve` command, and stores it inside the `x1` variable. `x1` is not displayed since the line ending prevents it.
3. Computes the solution to the system of equations $A*x=y$ by using matrix multiplication with the inverse of `A`, and stores it in the variable `x2`.
4. Displays the content of variable `x1` with the command `disp`.

Several comments are also present at the end of lines 2, 3 and 4 (after the `//`). These instructions need to be saved inside a text file, for example by using the SciNotes text editor. When saving, name the file and end it with the extension `.sce`, for example `testscript.sce`.

Tip › *Long instructions can also span several lines with the use of the command* `...` *This command can be inserted at the location that denotes the breaking point of the logical line which divides it into two distinct lines. This can be useful to improve script legibility and avoid writing very long command lines, for instance when calling a function with several input parameters. Similarly, to define a matrix, we can enter the coefficients by returning to a new line after each* `;` *or at any other point since in this case the use of* `..` *is optional. In all cases, the result will be the same regardless of the number of lines spanned by the command.*

```
// matrix defined across several lines
M= [ 1 2 3;
     4,5,6]
// instructions spanning several lines
sum(M,..
'c')
```

which returns:

```
-->//matrix defined across several lines

-->M= [ 1 2 3;
-->      4,5,6]
 M  =
    1.     2.     3.
    4.     5.     6.

-->//instructions spanning several lines

-->sum(M,..
-->'c')
 ans  =
    6.
    15.
```

Executing a script

Once a file is saved, there are several methods you can use to execute the commands it contains:

- from the SciNotes menu bar (or the equivalent keyboard shortcut):
  - EXECUTE > ...FILE WITH NO ECHO (Ctrl+Shift+E)
  - EXECUTE > ...FILE WITH ECHO (Ctrl+L)
  - EXECUTE > ...UNTIL THE CARET, WITH ECHO or, if you've selected part of the script EXECUTER > THE SELECTION WITH ECHO (Ctrl+E for both)

Figure 14.1 : Launching a script from the SciNotes menu bar

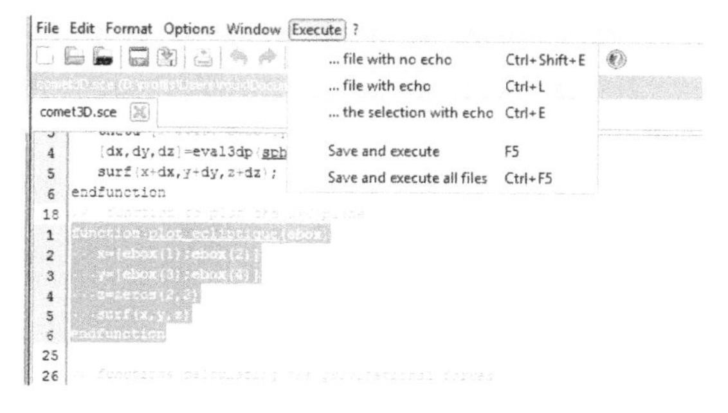

- from the following SciNotes toolbar icons:
  - EXECUTE
  - SAVE AND EXECUTE
  - SAVE AND EXECUTE ALL FILES

Figure 14.2 : Launching a script from the SciNotes toolbar

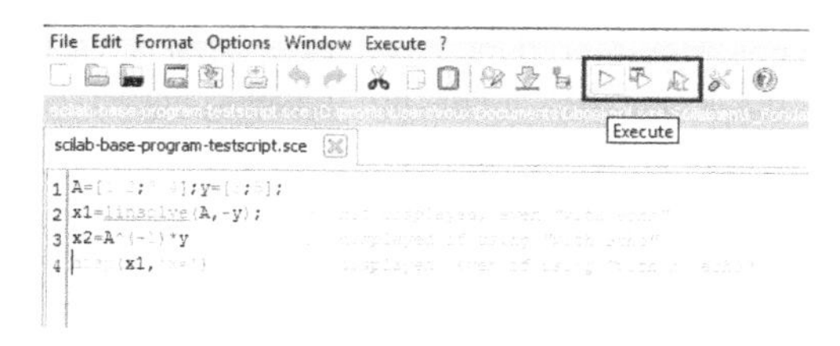

- from the Scilab console:
  - with a copy/paste of the script text into the console
  - with the menu FILE > EXECUTE (display without echo)
  - by using the exec command (display with or without echo according to the chosen option)

Figure 14.3 : Launching a script from the console menu bar

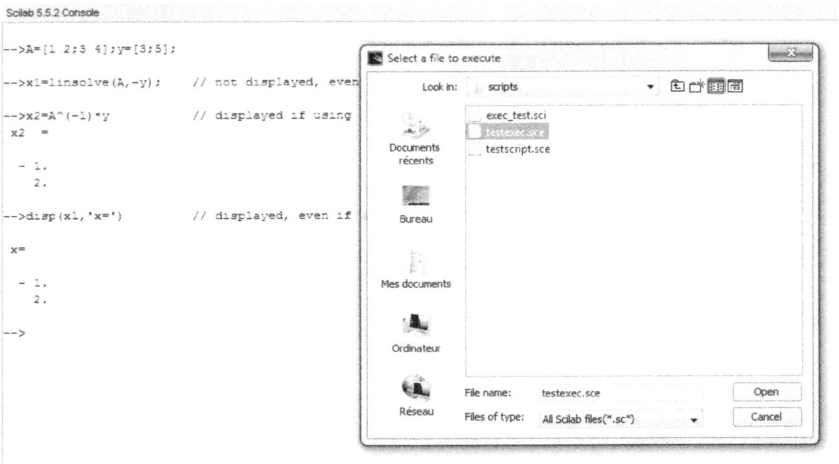

Setting the results display

Regardless of the method chosen to execute a script (out of the available options), Scilab executes the lines in the file one by one. However, during execution, what is displayed depends on the way in which the script is executed. If the chosen method is:

With echo

The commands along with their results are displayed in the console.

Without echo

The script is executed without console display.

You can also more precisely customize the way the results are displayed in the Scilab console by using the command exec which, according to the chosen option, leaves more or less space in the console between the commands and their results:

```
    0.

-->exec('testscript.sce',-1) // no echo, except for "disp"

 x=
```

```
  - 1.
    2.

-->exec('testscript.sce',0) // echo unless there is a ";"
 x2  =

  - 1.
    2.

 x=

  - 1.
    2.

-->// like option 0 + command prompt

-->exec('testscript.sce',1)
-->A=[1 2;3 4];y=[3;5];
-->x1=linsolve(A,-y);     // not displayed, even "with echo"
-->x2=A^(-1)*y            // displayed if using "with echo"
 x2  =

  - 1.
    2.
-->disp(x1,'x=')          // displayed, even if using "with no echo"

 x=

  - 1.
    2.

-->// like option 0 + new line

-->exec('testscript.sce',2)



 x2  =

  - 1.
    2.


 x=

  - 1.
    2.


-->// like option 1 + new line

-->exec('testscript.sce',3)

-->A=[1 2;3 4];y=[3;5];

-->x1=linsolve(A,-y);     // not displayed, even "with echo"
```

```
-->x2=A^(-1)*y              // displayed if using "with echo"
 x2  =

  - 1.
    2.

-->disp(x1,'x=')            // displayed, even if using "with no echo"

 x=

  - 1.
    2.
```

Note also that exec lets you execute the script interactively, meaning that it pauses after each command.

Finally, the mode command, like exec, makes it possible to specify the way in which the results are displayed in the console but this time, directly from the script code. It therefore performs the same way, regardless of the *option chosen for exec*. For instance, executing the script without echo:

```
mode(1)                   // execution mode with echo
A=[1 2;3 4];y=[3;5];
x1=linsolve(A,-y);        // not displayed, even "with echo"
x2=A^(-1)*y               // displayed if "with echo"
disp(x1,'x=')             // displayed even "with no echo"
```

returns:

```
-->mode(1)                   //execution mode with echo
-->A=[1 2;3 4];y=[3;5];
-->x1=linsolve(A,-y);        //not displayed, even "with echo"
-->x2=A^(-1)*y               //displayed if "with echo"
 x2  =
  - 1.
    2.
-->disp(x1,'x=')             //displayed even "with no echo"

 x=

  - 1.
    2.
```

Tip › *You can also execute part of a script with the following method:*

- *Use your mouse to select the area of interest in the text editor.*
- *Right-click to get to the popup menu.*
- *Select the sub-menu EVALUATE SELECTION WITH ECHO.*

Figure 14.4 : Launching a script from the SciNotes popup menu

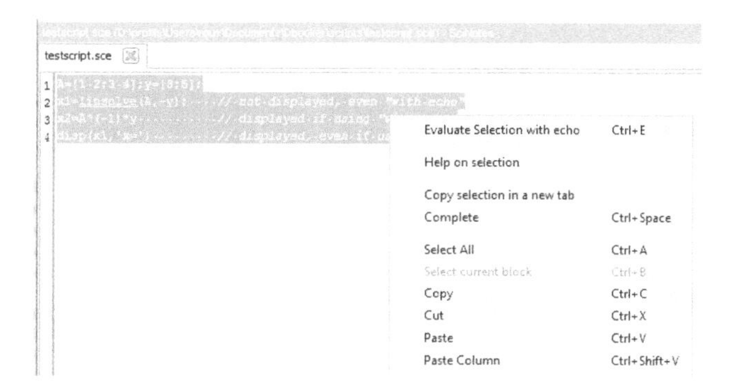

Caution › By default, executing a script from the SciNotes menu bar with the option WITH ECHO does not skip lines after each command display. This is contrary to what gets displayed when each line is entered in the console.

Tip › The commands `execstr` and `mgetl` also let you execute a script saved inside a file `file.sce`. The command `exec('file.sce')` is equivalent to `execstr(mgetl('file.sce'))`.

Halting the execution of a script

Sometimes it can be useful to interrupt a script while it is running to interact with the script user. A very common example is to ask the user if he wishes to carry on or not with the script execution. This can be achieved with the `input` command:

```
x=input('would you like to resume (1) or terminate (0) the script ?')
```

this displays the argument message in the console and waits for a value input (1 in this example), as shown below.

```
-->x=input('would you like to resume (1) or terminate (0) the
script ?')
 x  =
    1.
```

In this example, you retrieve a value entered from the keyboard and store it in a variable `x` straight from the Scilab console. This value can then be used to either resume or stop the script's execution. In other cases, it may be useful to pause the script to provide time for the user to analyze results and, potentially, decide to stop the computation. In order to do this, you can use the commands `pause` and `halt` which interrupt the script's execution. The execution is resumed according to the situation:

- After `pause`: the execution resumes as soon as the command `resume`, `return` or `quit` is entered.
- After `halt`: the execution resumes once the use hits the ENTER key.

To terminate the script's execution after pausing, enter the `abort` command.

For example:

```
x=1
pause
x=2
halt('hit ''Enter'' to resume')
x=3
```

returns:

```
-->x=1
 x  =
    1.

-->pause

-1->resume

-->x=2
 x  =
    2.

-->halt('hit ''Enter'' to resume')
hit 'Enter' to resume
-->x=3
 x  =
    3.
```

You'll notice that when coming across a `pause` command, the prompt displayed in the console is modified. It displays `-1->` (instead of the usual `-->`). This indicates that the execution is on hold. If other processes are launched and then halted, the prompt counter increments (`-2->`, then `-3->`, ...) indicating the total number of processes on hold. The prompt displays this as long as it is not instructed to `resume`, `return` or `abort`.

Caution › Entering **`resume`** *or* **`return`** *decrements by one the number of processes on hold whereas* **`abort`** *terminates all processes on hold.*

Tip › When a process is paused, the user retains control of the Scilab console. This means that you can use **`pause`** *to examine or even modify script variables during the execution. This option is very useful when tuning or debugging complicated scripts.*

*Caution › Do not confuse the **resume**, **return** and **quit** with the **exit** instruction which shuts down the active Scilab session!*

Scilab startup and shutdown scripts

Two important scripts located inside the `SCI/etc/` directory are automatically executed by Scilab:

- `scilab.start` during startup
- `scilab.quit` when shutting down

These two files manage the instructions that Scilab needs to execute during the software startup and shutdown (for example, it loads supplementary modules if you have some installed). If you wish to customize the way Scilab starts up for a given user, do not modify the `scilab.start` script. Instead, create a file inside the `SCIHOME` user directory called `.scilab` or `scilab.ini`. When Scilab starts up for the appropriate user, it automatically executes this file (if it exists) and performs operations that the user may not wish to repeat each time Scilab is launched. For example, if you want Scilab to always start up inside a given directory with a given location called `path`, add the line `cd 'path'` to your `scilab.ini` file which you will save inside the `SCIHOME` directory.

14.2. Dialog boxes

We have already covered how to display character strings in the console by using `disp` and `printf`, but sometimes it can be useful to display text in a popup window. This can be done with the command `messagebox`. This lets you display a window with:

- a message
- a title
- an icon (several images are available : error, warning, question mark...)
- buttons defined by a row matrix containing character strings (the names displayed on the buttons)

If a 5th optional parameter called `"modal"` is added (see Figure 14.5), then the window blocks access to the console. In other words, as long as the user hasn't closed the window or clicked one of its buttons, he/she won't be able to interact with the console.

*Tip › In **modal** mode, **messagebox** returns the button number (1,2...) that was clicked before regaining control of Scilab. If the window is closed without clicking a button, then **messagebox** returns 0.*

Figure 14.5 : Dialog box example

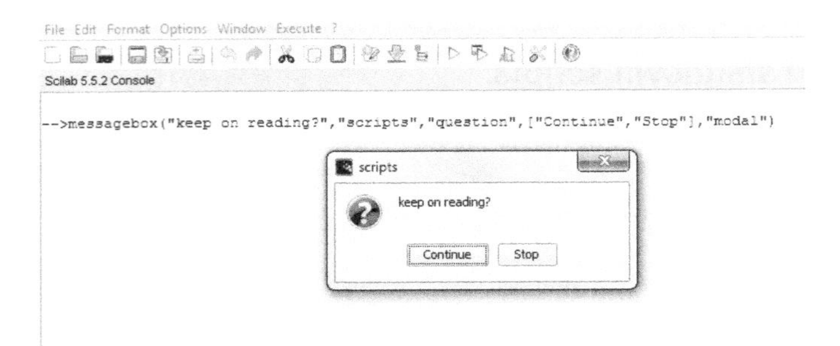

`messagebox` returns 1 if you click CONTINUE, 2 for STOP and 0 if the window gets closed.

Tip › The **`messagebox`** *text can get copied to the clipboard (use Ctrl+C).*

Other dialog boxes let you retrieve information entered in the console with the keyboard (such as constants, character strings...) while the script is running. We have already talked about `editvar` and `x_matrix` in chapter *Matrices*. You can also use:

`x_dialog`

Lets you retrieve values as a string of characters.

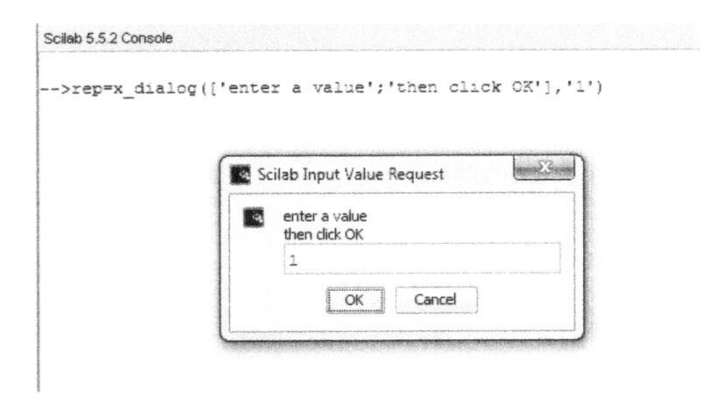

When OK gets clicked, `xdialog` retrieves the string entered (in this case 1) and stores it inside the variable `rep`.

```
-->rep=x_dialog(['enter a value';'then click OK'],'1')
 rep  =
 1
```

x_mdialog

Lets you retrieve several values as a strings matrix.

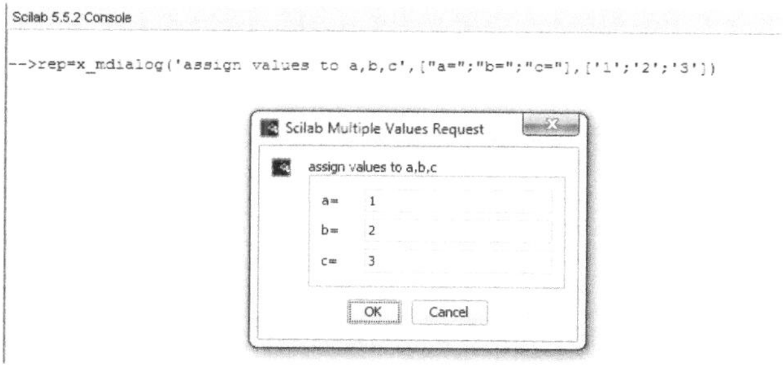

Scilab 5.5.2 Console

```
-->rep=x_mdialog('assign values to a,b,c',["a=";"b=";"c="],['1';'2';'3'])
```

Scilab Multiple Values Request

assign values to a,b,c

a= 1

b= 2

c= 3

OK Cancel

When one clicks OK, `xmdialog` retrieves a matrix made of strings `['1';'2';'3']` inside the variable `rep`.

```
-->rep=x_mdialog('assign values to a,b,c',["a=";"b=";"c="],
['1';'2';'3'])
 rep  =
!1  !
!   !
!2  !
!   !
!3  !
```

getvalue

Lets you retrieve several values while checking the type of the data retrieved.

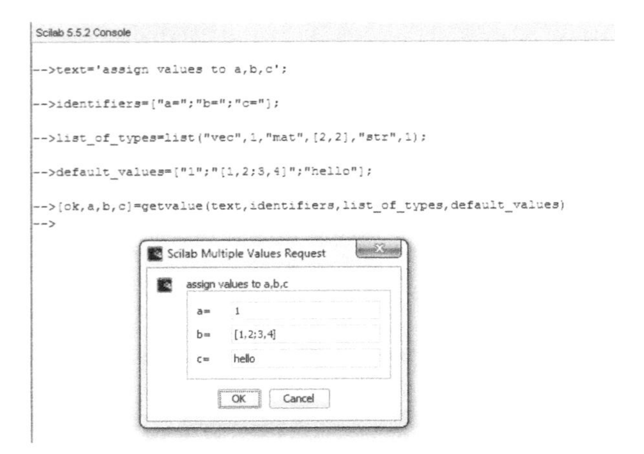

When one clicks OK, getvalue retrieves the values of the variables a, b, c and %T and stores it inside OK.

```
-->text='assign values to a,b,c';

-->identifiers=["a=";"b=";"c="];

-->list_of_types=list("vec",1,"mat",[2,2],"str",1);

-->default_values=["1";"[1,2;3,4]";"hello"];

--
>[ok,a,b,c]=getvalue(text,identifiers,list_of_types,default_values)
 c  =
 hello
 b  =
    1.    2.
    3.    4.
 a  =
    1.
 ok  =
  T
```

`x_choose` or `x_choose_modeless`

Lets you retrieve the index of an item that is double-clicked in a menu.

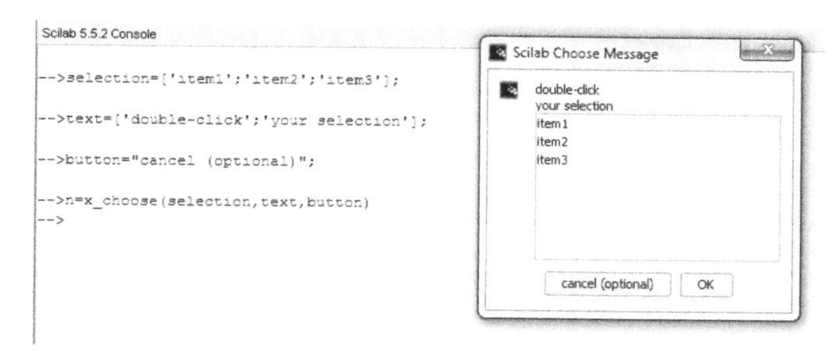

When one double-clicks ITEM2, `xchoose` retrieves and stores the value 2 inside variable `n`.

```
-->selection=['item1';'item2';'item3'];

-->text=['double-click';'your selection'];

-->button="cancel (optional)";

-->n=x_choose(selection,text,button)
 n  =
    0.
```

`x_choices`

Lets you select several choices by clicking on buttons.

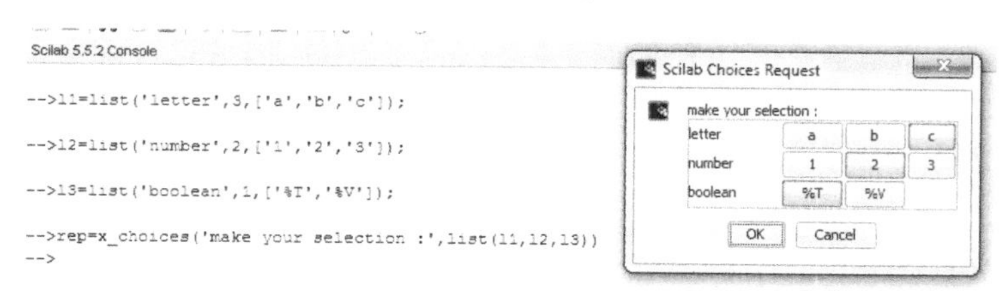

When one clicks on OK, `x_choices` retrieves the strings matrix `[3,2,1]` and stores it in `rep`.

```
-->l1=list('letter',3,['a','b','c']);

-->l2=list('number',2,['1','2','3']);
```

```
-->l3=list('boolean',1,['%T','%V']);

-->rep=x_choices('make your selection :',list(l1,l2,l3))
 rep  =
    3.    2.    1.
```

Tip › When the windows `x_dialog` ,`x_mdialog`, `getvalue` or `x_choices` are closed without entering or selecting any parameters, these functions return an empty matrix [ ]. On the other hand, `x_choose` returns the value 0.

Caution › The functions `x_dialog` and `x_mdialog` output strings. If you wish to retrieve numerical values, call the function `evstr` with strings as arguments.

You can use the following standard dialog boxes to manage files and paths:

- `uigetdir` to retrieve the path of a directory
- `uigetfile` to retrieve a file path
- `uiputfile` to save a file

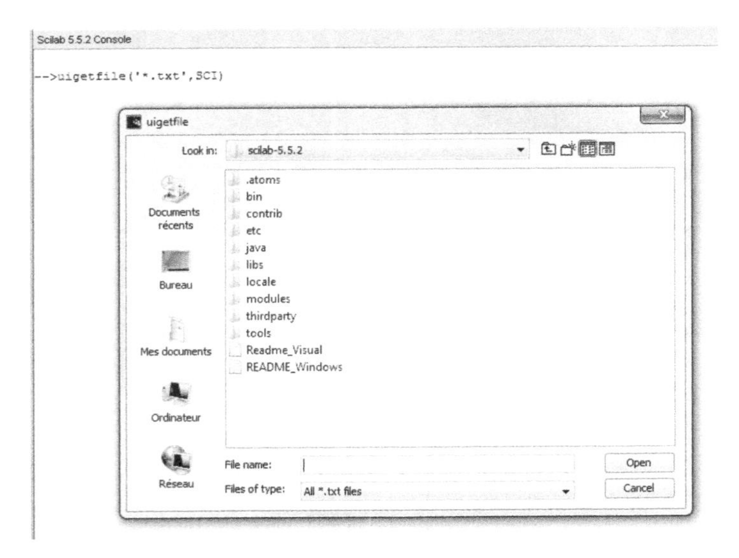

`uigetfile` dialog box to retrieve a file name: when you click on OK, the string `C:/Program Files/scilab-5.5.2/README_Windows.txt` is retrieved.

Caution › `uiputfile` does not save any data to a file, it only serves to confirm the saving process and retrieve the path of the file in which information will be saved. This information will then need to get passed to `mputl` or `write` or another function which then writes the data to the file (see Chapter Character Strings and Text Files).

15

Control Flow Statements

Control flow statements in a program let you control the sequence of execution of instructions and in some cases prevent the execution of certain commands. Scilab's language syntax lets you easily implement the more common constructs such as conditional branches and loops.

15.1. Conditional structures

Conditional structures use logic to test a condition. The result of this test determines which block of instructions is executed next. There are three types of conditional structures:

- `if`, `then`, `else`
- `select`, `case`
- `try`, `catch` (exception handling)

if, then, else

The most well-known structure is the *if-then-else* statement which uses the keywords `if`, `then`, `else`, `end`, as follows:

```
if condition then  instructions_1
             else  instructions_2
end
```

where:

- `condition` is a boolean expression (see Chapter *Booleans*)
- `instructions_1` and `instructions_2` are sets of Scilab instructions

For example, the following piece of code spells out (i.e. with a string stored inside the variable `txt`) if the variable `x` is equal to 1 or not.

```
x=1;
if x==1  then  txt='x is equal to 1';
    else txt='x is not equal to 1';
```

```
end
disp(txt)
x=2;
if x==1
    then  txt='x is equal to 1';
    else txt='x is not equal to 1';
end
disp(txt)
```

Which yields:

```
-->x=1;

-->if x==1  then  txt='x is equal to 1';
-->    else txt='x is not equal to 1';
-->end

-->disp(txt)

 x is equal to 1

-->x=2;

-->if x==1
-->    then  txt='x is equal to 1';
-->    else txt='x is not equal to 1';
-->end

-->disp(txt)

 x is not equal to 1
```

Tip › *If you wish to use multiple* if-then-else *statements, you can use the keyword* **`elseif`** *to simplify the code. In this instance, only one* **end** *is required to indicate the end of the block of instructions. On the other hand, two are needed when using* **`else if`** *as two words!*

```
-->//if then elseif

-->x=1;

-->if x==1  then  txt='x is equal to 1';
-->    elseif x==2 then txt='x is equal to 2';
-->    else txt='x is not equal to 1 nor 2';
-->end  //<-- only one "end"

-->disp(txt)

 x is equal to 1

-->//two nested if statements

-->x=2;
```

```
-->if x==1  then  txt='x is equal to 1';
-->    else if x==2 then txt='x is equal to 2';
-->         else txt='x is not equal to 1 nor 2';
-->         end //<-- "end" of second "if"
-->end //<-- "end" of first "if"

-->disp(txt)

 x is equal to 2
```

select, case

Another available structure is the *select-case* statement which uses the keywords `select`, `case`, `then`, `else`, `end`, as follows:

```
select expression
 case value_1 then
  instructions_1
 case value_2 then
  instructions_2
 ...
 case valuen then
  instructions_n
 else instructions
end
```

where:

- `expression` is a Scilab expression which can be evaluated (constant, boolean or other).
- `instructions_1` ... `instructions_n` are Scilab instructions.

The previous example written with the instruction `select-case` yields:

```
x=1;
select x
    case 1 then txt='x is equal to 1';
    else txt='x is not equal to 1';
end
disp(txt)
```

which, after execution, returns:

```
-->x=1;

-->select x
-->    case 1 then txt='x is equal to 1';
-->    else txt='x is not equal to 1';
-->end
```

```
-->disp(txt)

 x is equal to 1
```

Tip › In conditional structures, the **else** *instruction block is optional; it can be removed if it does not get used.*

```
-->x=1;

-->if x<>0 then txt='x is non-zero';
-->    // no else ...
-->end

-->disp(txt)

 x is non-zero
```

Caution › In all the different conditional structures we have described, the keyword **then***, which follows the keywords* **if** *and* **case***, is optional. However, we recommend not omitting it since it helps make the code easier to understand.*

```
-->// if then else

-->x=1;

-->if x==1   txt='x is equal to 1';
-->    else txt='x is not equal to 1';
-->end

-->disp(txt)

 x is equal to 1

-->// select

-->x=2;

-->select x
-->    case 1
-->           txt='x is equal to 1';
-->    else txt='x is not equal to 1';
-->end

-->disp(txt)

 x is not equal to 1
```

You may also notice that when **then** *is used, it needs to be placed on the same line as* **if** *and* **case***, otherwise you get an error (34) which states* Incorrect control instruction syntax.

```
-->x=1;
```

```
-->if x==1
-->    then txt='x is equal to 1'
  !--error 34
Incorrect control instruction syntax.
```

try, catch

Finally, the `try-catch` structure follows the format:

```
try instructions_1
     catch instructions_2
end
```

It makes it possible to keep track of errors that may appear while a command gets executed:

1. `instructions_1` gets executed.
2. `instructions_2` gets executed if and only if `instructions_1` returns and error (also see the command `errcatch`).

For example:

```
x=input('Compute 1/x for x=?');
try 1/x
catch disp('An error occurred!')
end
disp('The end of the script still gets executed.')
```

If you enter `x=0` when executing, it returns:

```
-->x=input('Compute 1/x for x=?');
Compute 1/x for x=?0
-->try 1/x
-->catch disp('An error occurred!')

 An error occurred!
-->end

-->disp('The end of the script still gets executed.')

 The end of the script still gets executed.
```

On the other hand with `x=2`, you get:

```
-->x=input('Compute 1/x for x=?');
Compute 1/x for x=?2
-->try 1/x
```

```
 ans  =
    0.5
-->catch disp('An error occurred!')
-->end

-->disp('The end of the script still gets executed.')

 The end of the script still gets executed.
```

15.2. Loops

You can create loops, which are control flow structures used to execute a block of instructions several times in succession. There are two types of loops:

- `while`
- `for`

while

The `while` loop is well-suited to situations where the number of repetitions is controlled by a logical condition. This structure is denoted in Scilab with the keyword `while`, as follows:

```
while condition
      instructions
end
```

where:

- `condition` is a boolean expression (see Chapter *Booleans*).
- `instructions` indicates a set of Scilab instructions.

For example:

```
txt='';
i=0;
while i<10
    i=i+1;
    txt=txt+string(i)+' ';
end
i,txt
```

yields:

```
-->txt='';

-->i=0;
```

```
-->while i<10
-->    i=i+1;
-->    txt=txt+string(i)+' ';
-->end

-->i,txt
 i  =
    10.
 txt  =
 1 2 3 4 5 6 7 8 9 10
```

Tip › To enhance code readability, you can add the keywords **do** *or* **then** *after the* **while condition** *block and before the Scilab instructions block. This is however not mandatory.*

```
-->txt='';

-->i=0;

-->while i<10 do
-->    i=i+1;
-->    txt=txt+string(i)+' ';
-->end

-->i,txt
 i  =
    10.
 txt  =
 1 2 3 4 5 6 7 8 9 10
```

Caution › If you use **do** *or* **then** *inside a* **while** *sequence, these keywords need to be placed on the same line as* **while**, *otherwise you will get an error (34) which states Incorrect control instruction syntax. (also see the note regarding the* **use of then inside conditional structures**).

```
-->txt='';
-->i=0;
-->while i<10
-->    do
-->    i=i+1;
-->    txt=txt+string(i)+' ';
-->end
              !--error 34
Incorrect control instruction syntax.
```

for

When the number of repetitions is known in advance, the for loop is better suited. This structure is denoted in Scilab as follows:

```
for var=vector
    instructions
```

```
end
```

where:

- var is a Scilab variable name.
- vector is a *row matrix* containing all the value that the var variable will take, consecutively, according to *the column order*.
- instructions indicates a set of Scilab instructions.

*Tip › In general, **var** adopts a set of numerical values stored inside **vector**, however a strings matrix or any other object type can also be used.*

For example:

```
txt='';
for i=1:10
    txt=txt+string(i)+' ';
end
i,txt
```

yields:

```
-->txt='';

-->for i=1:10
-->    txt=txt+string(i)+' ';
-->end

-->i,txt
 i  =
    10.
 txt  =
 1 2 3 4 5 6 7 8 9 10
```

The result is equivalent to what the while loop we previously studied returned. The counter i is set and managed by for.

*Tip › To create loops where the counter varies with a **step** other than 1, just construct a **vector** by adequately using the increment operator : such as **for i=a:step:b**.*

```
-->//increment with step=2

-->txt='';

-->for i=1:2:10
-->    txt=txt+string(i)+' ';
-->end

-->i,txt
```

```
 i  =
    9.
 txt  =
 1 3 5 7 9

-->//negative increment

-->txt='';

-->vector=[10:-1:1];

-->for i=vector
-->    txt=txt+string(i)+' ';
-->end

-->i,txt
 i  =
    1.
 txt  =
 10 9 8 7 6 5 4 3 2 1
```

If the `for` loop's indexing variable (which controls the number of iteration of the loop) doesn't exist before the loop, it gets created in the loop and remains afterwards. On the other hand, if the variable did exist prior to the `for` loop, then its value gets modified in the loop and will probably be different from its initial value at the end of the loop.

Caution › When `vector` *is not a row vector, the behavior of the* `for` *loop can be surprising! In this case, the variable* `var` *successively adopts the value of each matrix column:*

```
-->//row vector--> 3 iterations

-->vector=['one' 'two' 'three']
 vector  =
!one  two  three  !

-->for word=vector
-->    disp(word)
-->end

 one

 two

 three

-->//column vector--> only one iteration

-->vector=['one' 'two' 'three']'
 vector  =
!one    !
!       !
!two    !
!       !
!three  !
```

```
-->for mot=vector
-->    disp(word)
-->end

 three

-->// ordinary matrix--> one iteration per column

-->txt='';

-->vector=[1 2; 3 4]
 vector  =
    1.    2.
    3.    4.

-->for i=vector
-->    txt=txt+string(i)+' ';
-->end

-->i,txt
 i  =
    2.
    4.
 txt  =
!1 2   !
!      !
!3 4   !
```

Force a loop to continue or terminate

There exist two commands which force a loop to either continue or terminate (this is called an *unconditional jump*):

- `break` to get out of a loop that is running

```
-->//while loop

-->txt='';

-->i=0;

-->while %T
-->    i=i+1;
-->    txt=txt+string(i)+' ';
-->    if i>=10 then break
-->    end
-->end

-->i,txt
 i  =
    10.
 txt  =
```

```
 1 2 3 4 5 6 7 8 9 10

-->//for loop

-->txt='';

-->for i=1:10
-->    txt=txt+string(i)+' ';
-->    if i>=5 then break
-->    end
-->end

-->i,txt
 i  =
    5.
 txt  =
 1 2 3 4 5
```

- `continue` to skip straight to the next step in the loop that is running

```
-->//while loop

-->txt='';

-->i=0;

-->while i<10
-->    i=i+1;
-->    if (i>2)&(i<7) then continue
-->        else txt=txt+string(i)+' ';
-->    end
-->end

-->i,txt
 i  =
    10.
 txt  =
 1 2 7 8 9 10


-->//for loop

-->txt='';

-->for i=1:10
-->    if (i>2)&(i<7) then continue
-->        else txt=txt+string(i)+' ';
-->    end
-->end

-->i,txt
 i  =
    10.
 txt  =
 1 2 7 8 9 10
```

***Caution* › When dealing with two nested loops, the `break` command forces only one of the loops to stop:**

```
-->txt='';

-->i=0;

-->while i<2
-->    i=i+1;
-->    j=0;
-->    txt=txt+'i='+string(i)+', ';
-->    while j<5
-->        j=j+1;
-->        txt=txt+'j='+string(j)+', ';
-->        if j>2 then break  // get out at j=3
-->        end
-->    end
-->end

-->i,j,txt
 i  =
    2.
 j  =
    3.
 txt  =
 i=1, j=1, j=2, j=3, i=2, j=1, j=2, j=3,
```

***Tip* › Sometimes, the condition needed to get out of a loop may never be satisfied. In this case, the user will need to willingly stop the script's execution to regain control of Scilab. This can be done from the console:**

- *through the menu CONTROL/ABORT or CONTROL/INTERRUPT (see **Figure 15.1**)*
- *with the keyboard shortcut Ctrl+C which interrupts the loop*

When the loop gets interrupted, the user can examine the different variable values and decide:

- *to pursue the loop execution with* **`resume`** *or CONTROL/RESUME*
- *or stop the script execution with* **`abort`**

```
-->i=0;
-->while %T
-->    i=i+1;
-->end
-->//enter CTRL+C or from the menu bar "control->interrupt"
-->
Type 'resume' or 'abort' to return to standard level prompt.

-1->i
 i  =

    1739205.

-1->abort

-->i
```

```
 i  =

    1739205.
```

You will notice that each time the loop gets interrupted, a number appears next to the prompt **-1->**. *This number indicates how many loops are on hold, waiting to resume execution (also see* ***section Halting the execution of a script***). *It is not advisable to leave a large number of loops pending in a Scilab session.*

Figure 15.1 : *Stopping a loop from the* CONTROL *menu*

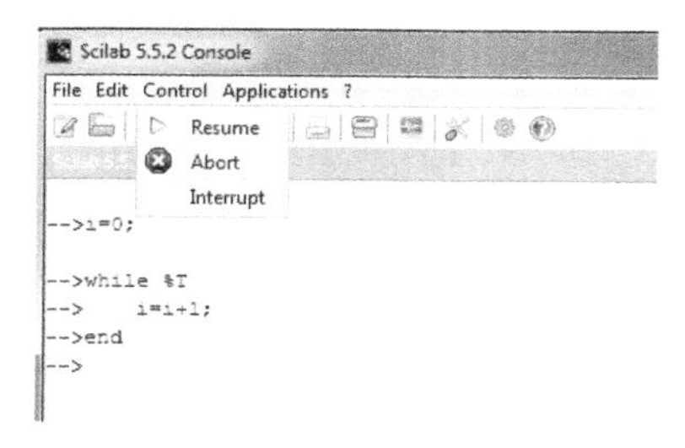

16
Functions

Throughout the different chapters, you have had a chance to notice the importance of native functions to perform standard operations, such as elementary mathematical functions. We will now cover how to create our own functions.

16.1. Defining a function

The syntax used to define a Scilab function is the following:

```
function [output1,output2,...]=function_name(input1,input2,...)
 instructions
endfunction
```

where:

- The keywords `function` and `endfunction` delimit the function definition.
- The first line (or *header*) defines the function's calling sequence (also see Section 3.1, *The online help*) with:
  - the output parameters `output1,output2,...` between `[ ]` and separated by commas
  - the equal sign `=` followed by the name of a function `function_name`
  - the input parameters `input1,input2,...` between `()` separated by commas
- The function body contains a list of `instructions` which lets you compute the output values `output1,output2,...` from the input values `input1,input2,...`

Take, for example, the case of a function `foo` which takes as input a number $1+x^2$. This function will be written in Scilab as:

```
function y=foo(x)
    y=1+x^2
endfunction
```

***Tip** › When there is only one output parameter, you can omit the [ ] around the output parameters in the function header.*

Then save the definition of a function in a text file with extension `sci` or `sce`. In order to use the function, *load* it in the Scilab environment the same way you would execute a script:

- from the text editor with the EXECUTE menu
- from the console with the `exec` command

Once the function is loaded in the current session, you can call it by following the calling sequence defined in the function.

```
-->function y=foo(x)
-->    y=1+x^2
-->endfunction


-->foo(2)  // returns 5
 ans  =
    5.
```

Caution › *It is strongly recommended not to enter the function body inside the Scilab console, but rather use the text editor and save the function definition inside a text file. When the* **`function`** *keyword is entered inside the Scilab console, it is impossible to regain control and perform any particular command before reaching the corresponding* **`endfunction`**.

```
-->function y=foo(x)
-->    y=1+x^2
-->    exp(1)          //not executed
-->    foo(1)          //not executed
-->
-->
-->
-->
-->endfunction       // lets you regain control!


-->
```

Tip › *A Scilab function can:*

- *have no input parameter. In this case, the function header's* **`(input1,input2,...)`** *part can be replaced by* **`()`** *or omitted*
- *have no output parameter. In this case the function header's* **`[output1,output2,...]=`** *can be replaced by* **`[]=`** *or omitted*

```
-->function y=foo1()
-->    y=1
-->endfunction
```

```
-->foo1()   // outputs 1
 ans  =
    1.


-->function foo2()
-->    disp('hello')
-->endfunction


-->foo2()    // doesn't output anything but displays 'hello'

 hello

-->a=foo2()   // no error

 hello

-->a          // but a isn't created

    !--error 4

 Undefined variable: a
```

A function that is loaded inside the working session is itself a variable belonging to the `function` type. It can therefore be copied to a different variable by using the assignment operator `=`, or used as argument of another function:

```
-->function y=foo(x)
-->    y=1+x^2
-->endfunction


-->typeof(foo)
 ans  =
 function

-->f=foo       // assign foo to f
 f  =
[y]=f(x)

-->f(2)       //outputs 5
 ans  =
    5.

-->typeof(f)
 ans  =
 function

-->disp(foo)  // displays the functions

[y]=foo(x)
```

You can retrieve the function's body in text format by using the command `fun2string`:

```
-->function y=foo(x)
-->    y=1+x^2
-->endfunction

-->fun2string(foo,'foo1')
 ans  =
!function y=foo1(x)  !
!                    !
!  y = 1 + x^2       !
!                    !
!endfunction         !
```

Similarly, with the `string` function, you can retrieve the input/output parameters and body of a function as separate strings:

```
-->function y=foo(x)
-->    y=1+x^2
-->endfunction

-->[out,in,txt]=string(foo)
 txt  =
!              !
!              !
!y = 1 + x^2   !
!              !
!              !
 in  =
 x
 out  =
 y
```

Caution › *Scilab's predefined functions are not all of the* **`function`** *type. Some of them belong to the* ***`fptr`*** *type:*

```
-->typeof(atan)   // arctangent function
 ans  =
 fptr

-->typeof(atanh)  // inverse hyperbolic tangent function
 ans  =
 function
```

Function of the type ***`fptr`*** *cannot be entered as arguments of other functions. Doing so returns an error with warning* error: Wrong type for argument #o: Function or string (external function) expected.

```
-->x=(1)./[-2 2 3];
```

```
-->typeof(atanh)
 ans  =
 function

-->atanh(x)
 ans  =
  - 0.549306144    0.549306144    0.346573590

-->feval(x,atanh)
 ans  =
  - 0.549306144    0.549306144    0.346573590

-->typeof(atan)
 ans  =
 fptr

-->atan(x)
 ans  =
  - 0.463647609    0.463647609    0.321750554

-->feval(x,atan) // error 211

    !--error 211

 feval: Wrong type for argument #2: Function or string (external
 function) expected.
```

Note › *You can also use the command* `deff` *to write functions:*

```
deff('[y]=mom_fonction(x)',instructions,options)
```

We also recognize the same arguments to define a function as those used with `function/endfunction`*:*

- *the line that defines the function's calling sequence*
- *a strings matrix containing the* `instructions` *to compute the output values* `output1,output2,...` *from the input values* `input1,input2,...`

The use of `deff` *is restricted to specific cases, for example when the commands that make up the body of the function are retrieved as strings. In all other cases, prioritize the use of the format* `function ... endfunction`*.*

```
-->deff('y=foo(x)','y=1+x^2')

-->foo(3)  // outputs 10
 ans  =
    10.
```

Tip › *You can also define a function inside another function with the keywords* `function` *and* `endfunction`*. For enhanced legibility, use the command* `deff`*.*

16.2. Function calling sequence

Contrary to the C language, Scilab functions can return several output parameters. The calling sequence to retrieve all the output parameters is outlined in function's first line (the line with `function`). For example, the function `foo` defined by:

```
function [a,b,c]=foo(x,y,z)
    a=x+y
    b=x*y
    c=z
endfunction
```

is called by following the calling sequence `[u,v,w]=foo(1,2,3)`. Note that it is possible to retrieve only part of the output parameters as long as it *follows the order in which they appear in the function definition*.

```
-->function [a,b,c]=foo(x,y,z)
-->    a=x+y
-->    b=x*y
-->    c=z
-->endfunction

-->foo(2,3,4)             // only the value of a is displayed
 ans  =
    5.

-->[a,b]=foo(2,3,4)       // a and b are retrieved as outputs
 b  =
    6.
 a  =
    5.

-->[a,b,c]=foo(2,3,4)  // a,b,c are retrieved as outputs
 c  =
    4.
 b  =
    6.
 a  =
    5.
```

Caution › Recall (see ***Section 2.2, Using the console****) that the last result of a command gets stored inside the* **ans** *variable if it isn't stored inside a specific variable. This is what happens when the function results are not stored inside variables.*

To define functions with optional input/output parameters, you can use the following keywords:

- `varargin` which will be the last input argument

- varargout which will be the last output argument

varargin and varargout are viewed as lists and manipulated as such inside the function body.

```
-->function [y,varargout]=foo(x,varargin)
-->    y=x^2
-->// number of optional input arguments
-->    n=length(varargin)
-->    if n>0 then
-->        for i=1:n
-->            printf('optional argument number %d :',i)
-->            disp(varargin(i))
-->//use optional output arguments
-->            varargout(i)=varargin(i)
-->        end
-->    end
-->endfunction


-->// no optional arguments

-->foo(1)
 ans  =
    1.

-->// 3 optional input arguments

-->foo(1,2,3,4)
optional argument number 1 :
    2.
optional argument number 2 :
    3.
optional argument number 3 :
    4.
 ans  =
    1.

-->// retrieve 2 optional output arguments

-->[y,a,b]=foo(1,2,3,4)
optional argument number 1 :
    2.
optional argument number 2 :
    3.
optional argument number 3 :
    4.
 b  =
    3.
 a  =
    2.
 y  =
    1.
```

You can also use argn to determine how many input/output arguments were used during the function call:

```
-->function [varargout]=foo(varargin)
-->    [lhs,rhs]=argn()  //  number of input/output arguments
-->    printf('there are %d input arguments\n',rhs)
-->    printf('there are %d output arguments\n',lhs)
-->    for i=1:lhs
-->        varargout(i)=i
-->    end
-->endfunction


-->foo(1,2,3)
there are 3 input arguments
there are 1 output arguments
 ans  =
    1.

-->[a,b,c]=foo(1,2)
there are 2 input arguments
there are 3 output arguments
 c  =
    3.
 b  =
    2.
 a  =
    1.
```

Tip › In the specific case of functions that only take strings as input, it is tedious to use apostrophes ' ' around each argument. In this case, there exists a simpler calling sequence:

```
function_name  input1 input2
```

For example:

```
-->function y=foo(a,b,c)
-->    y='input parameters :'+a+','+b+','+c
-->endfunction

-->foo('1','abc','SCI')
 ans  =
 input parameters :1,abc,SCI

-->foo 1 abc SCI
 ans  =
 input parameters :1,abc,SCI
```

This syntax can be applied to certain functions such as **cd** *or* **ls** *as long as the arguments do not contain any Scilab special characters.*

```
-->cd SCI       // here SCI is interpreted as a Scilab variable!
 ans  =
 D:\profils\Users\roux\AppData\Local\scilab-5.5.2

-->ls *.txt     // *.txt is viewed as '*.txt'
 ans  =
!README_Windows.txt  !
!                    !
!Readme_Visual.txt   !

-->cd contrib  // contrib  is interpreted as 'contrib'
 ans  =
 D:\profils\Users\roux\AppData\Local\scilab-5.5.2\contrib

-->cd ..        // waiting for end of command
-->             // equivalent to 'cd home'
 ans  =
 D:\profils\Users\roux

-->home
 home  =
 D:\profils\Users\roux
```

16.3. Scope of variables and arguments

Scilab's function execution mecanism follows the same principles as for script languages. In other words, when a function gets called:

1. A new instance of Scilab opens up in which the input parameters are copied.
2. The function body's instructions are executed inside this new Scilab instance.
3. The Scilab instance closes and the output parameters are returned to the initial Scilab instance in which the function was called.

Specifically, variables created inside the function body only exist inside the Scilab instance linked to the function call. These variables do not conflict with variables which may have the same name in the function calling environment. We call these *local variables*.

```
-->function y=foo(x)
-->    disp('dans ''foo''  x='+string(x))
-->    y=1+x^2
-->    disp('dans ''foo''  y='+string(y))
-->endfunction

-->x=2
 x  =
    2.
```

```
-->z=foo(1)  //inside "foo" x is 1
 dans 'foo'  x=1
 dans 'foo'  y=2
 z  =
    2.

-->x         // x is still equal to 2
 ans  =
    2.

-->y         // y  doesn't exist here
!--error 4
Undefined variable: y
```

When a variable gets called inside a function body in which it is not defined, Scilab searches if a variable of the same name exists inside the environment in which the function was called. It then takes its value by default.

```
-->function y=foo(x)
-->    disp('inside ''foo''    x='+string(x))
-->    disp('at the beginning      a='+string(a))
-->    y=a+x^2
-->    a=a+1   // modifies a locally
-->    disp('at the end      a='+string(a))
-->endfunction


-->a=1
 a  =
    1.

-->x=2
 x  =
    2.

-->foo(3)  //inside "foo" x is 3  and a gets modified

 inside 'foo'    x=3

 at the beginning      a=1

 at the end      a=2
 ans  =
    10.

-->x        // x is still  2
 ans  =
    2.

-->a        // a is still 1
 ans  =
    1.
```

Caution › *If a variable appears inside a function body and isn't defined anywhere, executing the function produces an error (4) with warning* Undefined variable:

```
-->function y=foo(x)
-->    y=z+x^2
-->endfunction


-->clear z   // z doesn't exist

-->foo(1)    // error

    !--error 4

 Undefined variable: z
```

By default, the modifications performed on a variable inside the function body do not carry over to the environment in which the function was called. We can change that behavior by using the command `global`. This lets us declare a global variable inside the current environment and within the function body. For such a variable, changes inside the function body leads to changes inside the calling environment.

```
-->a=10;b=100;

-->global a b;    // declare global variables

-->function y=foo(x)
-->    global a;  // only a is global
-->    disp('dans ''foo''    a='+string(a))
-->    y=a+x^2
-->    a=2*a      // modifies a globally
-->    b=2*b      // b is locally modified
-->    disp('dans ''foo''    b='+string(b))
-->endfunction


-->foo(1)       // inside "foo" a is 10

 dans 'foo'    a=10

 dans 'foo'    b=200
 ans  =
    11.

-->a            // a gets modified outside
 ans  =
    20.

-->b            // b does not get modified
 ans  =
    100.
```

With the command `macrovar`, it is possible to retrieve all the variables that appear inside a function's body and organize them inside a list of strings:

- input/output variables that are used inside the function calling sequence
- external variables and functions called inside the function body
- local variables solely defined inside the function body

```
-->function [a,b]=foo(x,y)
-->    u=x+y         // u is a local variable
-->    v=x*y         // v is a local variable
-->    a=u+c         // c is an external variable
-->    b=u*exp(v)  // exp is an external functions
-->endfunction


-->L=macrovar(foo)   //   L=list(in,out,nolocal,called,local)
 L  =

       L(1)

!x  !
!   !
!y  !

       L(2)

!a  !
!   !
!b  !

       L(3)

 c

       L(4)

 exp

       L(5)

!u  !
!   !
!v  !
```

Tip › *Scilab's flexibility regarding variable scope makes it very easy to write recursive functions. For example, recursively computing a factorial is coded as:*

```
-->function y=fact(n)
-->     if n==0 then y=1
-->     else disp(' beginning of call to fact('+string(n-1)+')')
-->         y=n*fact(n-1)
-->     end
-->     disp(' end of call to fact('+string(n)+')')
-->endfunction


-->fact(3)

  beginning of call to fact(2)

  beginning of call to fact(1)

  beginning of call to fact(0)

  end of call to fact(0)

  end of call to fact(1)

  end of call to fact(2)

  end of call to fact(3)
 ans  =
    6.
```

Caution › *In Scilab, the factorial computation is performed by default with the function* **`factorial`**.

17

Advanced Programming

To supplement this presentation of programming with Scilab, here are a few tips to help optimize your approach to programming.

17.1. Error handling

As the program gets more complex, the risk of performing errors increases. We have already talked about some common errors inside Part *Computing* and Part *Programming*. The presence of errors gets signaled inside the console through messages. You need to be able to understand these messages well in order to handle these exceptions. Scilab's error messages are always associated to a number. A list can be found at the `error_table` help page (enter `help error_table` inside the console) [see Figure 17.1].

Figure 17.1 : *List of Scilab errors*

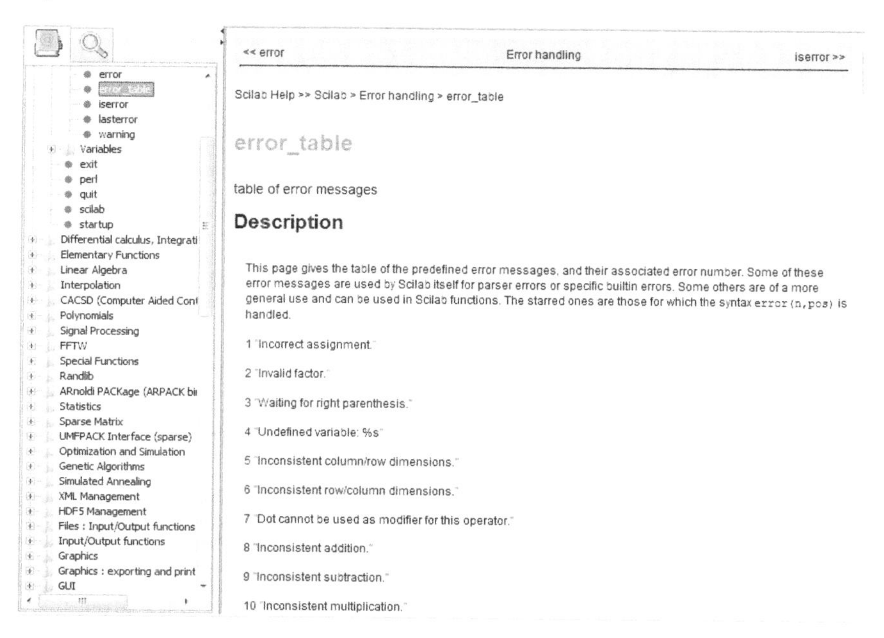

When you create your own Scilab functions, you can also create custom exceptions along with their own error number, by using the command `error`. Similarly, you can also display a warning in certain cases by using the command `warning`.

```
-->function y=foo(x)
-->    if x==0 then                    // 1/0 is undefined
-->        error('x cannot be 0!',9999)
-->    // 1/x too small for a good precision
-->    elseif abs(x)>1D15 then
-->        y=(1+1/x)^x;
-->        warning(['computation is too imprecise since x is too
 big!';..
-->        ' usually foo(x) ~ e=2.7182818...'])
-->    else y=(1+1/x)^x            // normal case
-->    end
-->endfunction

-->foo(1D10)    // normal computation
 ans  =
    2.7182821

-->foo(1D100)   // warning
WARNING: computation is too imprecise since x is too big!
          usually foo(x) ~ e=2.7182818...

 ans  =
    1.

-->foo(0)       // error 9999

    !--error 9999

 x cannot be 0!
```

Caution › *Do not mix up an error and a warning:*

- *When Scilab returns an error, the function call is not correctly terminated and no result is returned in the output variables.*
- *When Scilab returns a warning, the function call ends properly and the results are returned as outputs.*

With the `return` command, you can stop the function execution and return the state of certain variables inside the Scilab instance that called the function. This includes variables that do not belong to the function's output parameters:

```
-->function foo(x)
-->    txt_foo='x is equal to '+string(x)
-->    txt_sci=return(txt_foo)
-->    disp('what follows doesn''t get executed')
-->endfunction
```

```
-->foo(1)

-->txt_sci    // the txt_sci variable does exist here
 ans  =
 x is equal to 1

-->txt_foo    // the txt_foo variable doesn't exist here

    !--error 4

 Undefined variable: txt_foo
```

Caution › *Without argument,* **`return`** *is equivalent to* **`resume`**. *Specifically, and as opposed to the C language syntax, the command* **`return z`** *does not terminate the function execution by returning the value of* `z`*:*

```
-->function y=foo(x)
-->    z=1+x^2
-->    return z
-->endfunction

-->clear y  // no y variable

-->foo(0)   // the execution isn't successful

    !--error 4

 Undefined variable: y
```

As you can see from the function header, Scilab tries to retrieve the result of the function **`foo`** *inside the variable* `y`*. If you change the function header to* **`function foo(x)`**, *it will return a result inside the variable* `z`.

You can automatically retrieve information about a script or function error by using the command `lasterror`. If the error originates from a script executed with the commands `exec` and `execstr`, use the option `errcatch` to ensure the error gets captured by `lasterror`.

```
-->function foo()
-->    1/0
-->endfunction

-->// error inside a function

-->ierr=execstr('foo()','errcatch','m')
 !--error 27
Division by zero...

 ierr  =
    27.
```

```
-->[str,n,line,func]=lasterror()  //  error inside func="foo"
 func  =
 foo
 line  =
    2.
 n  =
    27.
 str  =
 Division by zero...

-->// error inside a command

-->ierr=execstr(['a=2';'b=1';'1/0'],'errcatch','m')
1/0
   !--error 27
Division by zero...

 ierr  =
    27.

-->[str,n,line,func]=lasterror()  //  here func=""
 func  =

 line  =
    3.
 n  =
    27.
 str  =
 Division by zero...
```

Caution › When an error is produced inside a script, the code that follows does not get executed unless the error is produced in:

- *the command* `try-catch`
- *the command* `exec` *or* `execstr` *called with the option* `errcatch`

To identify the source of an error, you can use the command `where` which lets you retrieve the calling sequence that lets you execute one script line.

```
-->function y=foo1(x)
-->    y=foo2(x)
-->endfunction


-->function y=foo2(x)
-->    y=1+x^2
-->    [line,fonc]=where()
-->    txt='function <'+fonc+'> at line '+string(line)
-->    disp(txt)
-->endfunction


-->foo1(2)
```

```
!function <foo2> at line 3          !
!                                   !
!function <foo1> at line 2          !
!                                   !
!                                   !
 ans  =
    5.

-->foo2(2)

!function <foo2> at line 3          !
!                                   !
!                                   !
 ans  =
    5.
```

You can also use the command `whereami`:

```
-->function y=foo1(x)
-->     y=foo2(x)
-->endfunction


-->function y=foo2(x)
-->     y=1+x^2
-->     whereami()
-->endfunction


-->foo1(2)
whereami   called at line 3 of macro foo2
foo2       called at line 2 of macro foo1
 ans  =
    5.

-->foo2(2)
whereami   called at line 3 of macro foo2
 ans  =
    5.
```

Caution › *When you are stuck inside a function's script or loop, you can interrupt the script execution (pause mode) and use* **`where`** *or* **`whereami`** *to identify which part of the script is causing issues.*

When you load a file that contains a function definition with the command `exec`, you may notice a warning *Warning : Redefining function* displayed in the console. It signals that the function's definition has changed. You can modify this behavior by using `funcprot` to remove the warning or replace it with error 111.

```
-->function y=foo(x)
-->    y=1+x^2
-->endfunction

-->funcprot(0)
 ans  =
    1.

-->function y=foo(x)
-->    y=2+x^2
-->endfunction    // no warning

-->// retrieve the current value of "funcprot" :

-->previousprot=funcprot()
 previousprot  =
    0.

-->// change and retrieve the current value of "funcprot" :

-->previousprot=funcprot(1)
 previousprot  =
    0.

-->function y=foo(x)
-->    y=3+x^2
-->endfunction    // warning (default behavior)
Warning : redefining function: foo                     . Use funcprot(0)
 to avoid this message


-->funcprot(2)    // change the current value of "funcprot"
 ans  =
    1.

-->function y=foo(x)
-->    y=4+x^2
-->endfunction    // produces error  111

    !--error 111

 Trying to re-define function foo.
```

Caution › *As with other variable types, you can individually protect variables containing functions with the command* **`predef`** *(see* **Section 8.3, Advanced variable management***).*

17.2. Function optimization

Scilab functions are compiled functions by default. They get compiled when loaded so as to optimize their execution when they get called. If you'd like to get some information on the execution time of each command in a function body, you need to first *profile* it:

- by using the command `add_profiling`
- by using the option `'p'` if the function is defined with the command `deff`

You can then execute the function and retrieve the information with the commands `profile`, `showprofile`, `plotprofile`.

```
-->function x=foo(n)
-->    k=1:n
-->    x=sum(k.^(-2))
-->endfunction


-->add_profiling('foo')    // profiled function
Warning : redefining function: foo                    . Use funcprot(0)
 to avoid this message


-->foo(1D5)                // execute foo
 ans  =
    1.644924067

-->profile(foo)      // retrieve the execution profile's matrix
 ans  =
    1.    0.       0.
    1.    0.001    4.
    1.    0.005    6.
    1.    0.       0.

-->showprofile(foo) // execution profile as character strings
|1|0    |0| 1: function x=fun(n)
|1|0    |4| 2:    k = 1:n
|1|0.01|6| 3:    x = sum(k.^(-2))
|1|0    |0| 4: endfunction

-->plotprofile(foo) // execution profile in graphics form
```

In this example, when `foo(1D5)`gets executed, the third command in the function requires the most CPU time.

Figure 17.2 : Graphs created with `plotprofile`

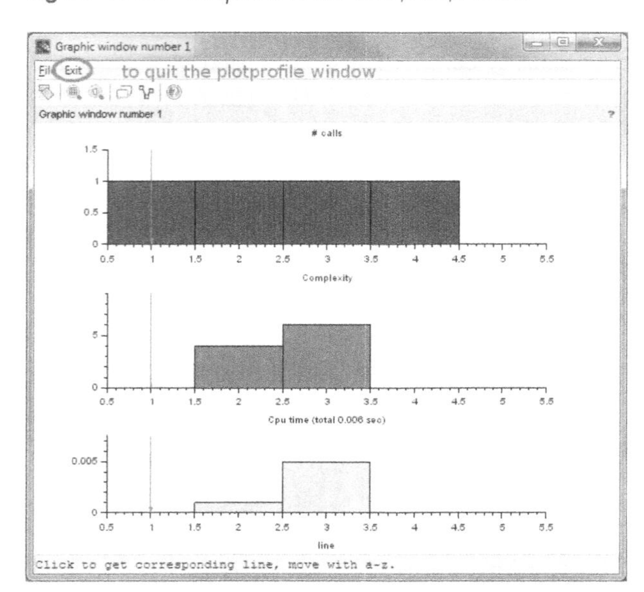

*Caution › `plotprofile` launches a graphics window (see **Figure** 17.2). You can only regain control in Scilab once this window is closed by clicking the EXIT button in the graphics window toolbar. If you close this window any other way, Scilab won't respond until you close the session.*

To improve the execution speed of scripts and functions, recall that many loops can be rewritten as matrix operations which execute much more rapidly. You can confirm this with the use of the commands `tic` and `toc`:

```
-->//loop calculation time

-->tic();

-->S=0;

-->for i=1:10^6
-->    S=S+i;
-->end

-->time=toc()
 time  =
    1.146

-->// matrix operation calculation time
```

```
-->tic();

-->S=sum(1:10^6);

-->time=toc()
 time  =
    0.009
```

Likewise with recursive functions, matrix operations help us avoid performing several recursive calls with one command. You can see this in this Fibonacci suite calculation:

```
-->//bad recursion

-->function y=Fibonacci1(n)
-->    if n<=1 then y=1
-->        else y=Fibonacci1(n-1)+Fibonacci1(n-2)
-->    end
-->endfunction


-->tic()

-->Fibonacci1(25)
 ans  =
    121393.

-->time=toc()
 time  =
    0.759

-->//matrix structure use

-->function y=Fibonacci2(n)
-->    F=[0 1; 1 1]
-->    u=(F^n)*[1;1]
-->    y=u(1)
-->endfunction


-->tic()

-->Fibonacci2(25)
 ans  =
    121393.

-->time=toc()
 time  =
    0.003
```

Caution › *When a recursive function call generates more than one call to itself, the function call count increments exponentially which quickly saturates the memory. In this case an error (26) gets returned which states* Too complex recursion! (recursion tables are full).

17.3. Object-oriented programming

To improve the programming quality, it is often of interest to create one's own object types. The principles of object-oriented programming can be established with Scilab by using typed lists :

- You can create an instance with the command `mlist` by following the format `instance=mlist(['new_type','field1',...,'fieldn'],value1,...,valuen)`.
- You can then access the different fields of an instance in the same way as for structures: `instance('field1')` or with the extraction operator "`.`" by entering `instance.field1`.

Take the example of the `point` type which is defined to represent a point of the plane with two coordinates `x=1` and `y=0`. To create an instance `M` of this type, one writes `M=mlist(['point','x','y'],1,0)`.

```
-->M=mlist(['point','x','y'],0,1)  // creates the point M
 M  =
        M(1)

!point  x  y  !

       M(2)
    0.

       M(3)
    1.

-->typeof(M)                       // M belongs to the point type
 ans  =
 point

-->M.x                             // coordinate x
 ans  =
    0.

-->M.y                             // coordinate y
 ans  =
    1.
```

Tip › For greater clarity and simplicity, it is recommended to create a function to create instances from each field's values. Take for example the function ***`make_point`****:*

```
function M=make_point(x,y)
    M=mlist(['point','x','y'],x,y)
endfunction
```

Similarly, to improve error handling in your previous work, you can create a function which checks the integrity of the data for a given instance. For example, the function **check_point**:

```
function check_point(M)
  if typeof(M)<>'point' then  whereami()
    error(10001,'the argument is not of the ''point'' type !')
  elseif length(M)<>3 then whereami()
    error(10002,'the point does not have two coordinates!')
  elseif (type(M.x)<>1)|(type(M.y)<>1) then whereami()
    error(10003,'the point''s coordinates are not numbers!')
  elseif (length(M.x)<>1)|(length(M.y)<>1) then whereami()
    error(10004,'the point''s coordinates are number tables!')
  elseif (imag(M.x)<>0)|(imag(M.y)<>0) then whereami()
    error(10004,'the point''s coordinates are not real!')
  end
endfunction
```

When the point's definition is correct, the function doesn't have any effect. In other cases, it returns an error message and lets you track the error's origin with the help of a **whereami**.

```
-->check_point(M)        // M is indeed a point

-->N=make_point(0,%i);   // point is not valid

-->check_point(N)        // N  is not a point
whereami    called at line 10 of macro check_point

    !--error 10004

 the point's coordinates are not real!
```

You can then define functions that call variable of this new type:

```
-->function l=distance(A,B)
-->    check_point(A)
-->    check_point(B)
-->    l=sqrt((A.x-B.x)^2+(A.y-B.y)^2)
-->endfunction


-->A=make_point(0,1)
 A  =

       A(1)

!point  x  y  !

       A(2)

    0.
```

```
        A(3)

    1.

-->B=make_point(1,0)
 B  =

        B(1)

!point  x  y  !

        B(2)

    1.

        B(3)

    0.

-->distance(A,B)
 ans  =
    1.414213562
```

More importantly, you can redefine the behavior of common operators (`+`, `-`, `*`, `^`, `/`, `\`, `.`, `()`, `[]`, `<`, `>`, `==`, `...`) for this new data type by using operator overloading. To extend an operator `op` (which can be chosen from the list `+`, `-`, `*`, `^`, `/`, `\`, `.`, `()`, `[]`, `<`, `>`, `==`, `...`), you need to define a function called:

- `%type1_op_type2` for binary operators
- `%type_op` or `%op_type` for unary operators

where `type`, `type1` and `type2` are codes that symbolize the different data types. The list of codes for operators and predefined types in Scilab is accessible from the online help (type `help overloading` in the console) [see Figure 17.3]. For example:

- `a` designates the operator `+`
- `m` designates the operator `*`
- `s` designates the operator `-`
- ...
- `s` designates the type `constant` (scalars)
- `c` designates the type `string`
- `b` designates the type `boolean`
- ...

Figure 17.3 : Online help page for `overloading`

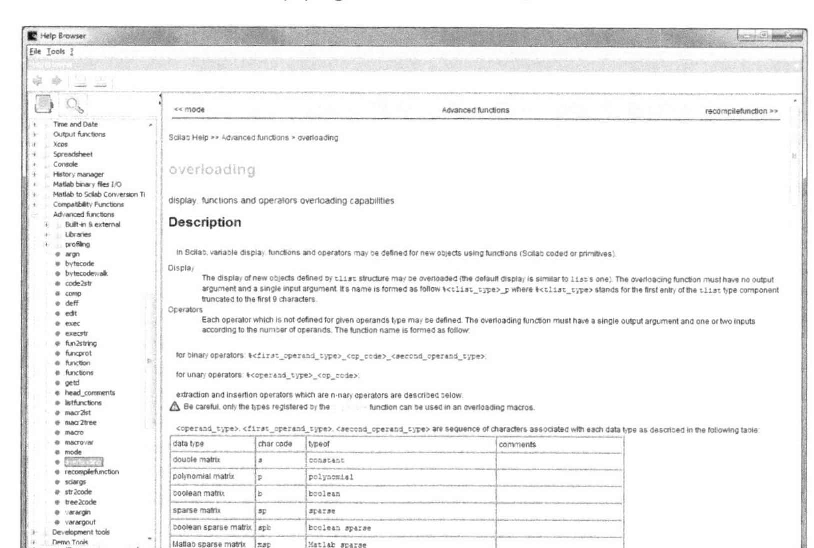

Tip › *First and foremost, overloading lets you redefine the way in which type's instances are displayed in the Scilab console. For this purpose, you must define (or modify) the function* **`%newtype_p`**. *For example, in order to modify the display of objects of the* **`point`** *type seen above, you must write a function* **`%point_p`** *which can then use commands such as* **`disp`** *or* **`printf`**:

```
-->M=make_point(0,1)  // creates point M
 M  =

         M(1)

!point  x  y  !

         M(2)

      0.

         M(3)

      1.

-->function %point_p(P)
-->     check_point(P)
-->     printf('(x=%f,y=%f)\n',P.x,P.y)
-->endfunction
```

```
-->M //  point M's display is modified
 ans  =
(x=0.000000,y=1.000000)
```

Example 17.1 : *Redefining the operator * for a user type*

To redefine the operator * such that O*P is the symmetry of point P with respect to point O, you must proceed as follows:

```
-->function M=%point_m_point(O,P)
-->    check_point(P)
-->    check_point(O)
-->    M=make_point(0,0)
-->    M.x=O.x-(P.x-O.x)
-->    M.y=O.y-(P.y-O.y)
-->endfunction

-->O=make_point(0,1)  // creates point O
 O  =
(x=0.000000,y=1.000000)

-->P=make_point(1,0)  // creates point P
 P  =
(x=1.000000,y=0.000000)

-->M=O*P                  // M=symmetric of P with respect to O
 M  =
(x=-1.000000,y=2.000000)
```

Caution › *When you use a usual operator (+, -, *, /, ...) on a data type for which it is not defined, it returns an error (144) which warns* Undefined operation for the given operands.

```
-->x=1     // a real number
 x  =
    1.

-->s='2'  //a string
 s  =
 2

-->x+s     // error since %s_a_c is undefined

    !--error 144

Undefined operation for the given operands.
check or define function %s_a_c for overloading.
```

You may notice that the message indicates the name of the function which needs to be defined to overload the operator. In the previous example, it involves %s_a_c *where:*

- a *designates the operator +*

- `s` *designates the type* `constant` *(scalars)*
- `c` *designates the type* `string`

```
-->function txt=%s_a_c(s,c)
-->    txt=string(s)+c
-->endfunction


-->x=1    // a real number
 x  =
    1.

-->s='2'  //a string
 s  =
 2

-->x+s    // %s_a_c returns the string  '12'
 ans  =
 12

-->s+x    // error since %c_a_s is undefined

    !--error 144

Undefined operation for the given operands.
check or define function %c_a_s for overloading.
```

17.4. Documenting your functions

You can write your own online help for functions you developed. The Scilab documentation must be written in DocBook format (see `help 'Scilab documentation format'` for more details). For each function `foo`, you must write a file `foo.xml` containing its description. Several functions let you generate templates for these files:

- `help_skeleton` lets you create a blank template.
- `manedit` let the Scilab text editor open a blank template.
- `help_from_sci` lets you retrieve the comments that follow the function header.

Once the help files are written, incorporate them to the Scilab online help with the commands:

- `xmltojar` to convert `xml` files to the `jar` format (other formats can also be created: `html` with `xmltohtml`, `pdf` with `xmltopdf`...)
- `add_help_chapter` to add a chapter to the online help and `del_help_chapter` to delete it.

For example:

```
-->function y=foo(x)
-->    //header comments
-->    //y,x= two real numbers
-->    y=1+x^2
-->    //closing comments
-->endfunction

-->head_comments('foo')
function [y] = foo(x)
header comments
y,x= two real numbers

-->txt=help_skeleton('foo');

-->mputl(txt,'foo.xml');
```

yields the help file:

```
<?xml version="1.0" encoding="UTF-8"?>
<!--
 * Add some comments about XML file
-->
<refentry xmlns="http://docbook.org/ns/docbook" xmlns:xlink="http://
www.w3.org/1999/xlink" xmlns:svg="http://www.w3.org/2000/svg"
 xmlns:mml="http://www.w3.org/1998/Math/MathML" xmlns:db="http://
```

```
docbook.org/ns/docbook" version="5.0-subset Scilab" xml:lang="fr_FR"
 xml:id="foo">
  <info>
    <pubdate>$LastChangedDate: 09-12-2012 $</pubdate>
  </info>
  <refnamediv>
    <refname>foo</refname>
    <refpurpose>Add short description here. </refpurpose>
  </refnamediv>
  <refsynopsisdiv>
    <title>Calling Sequence</title>
    <synopsis>y = foo(x)</synopsis>
  </refsynopsisdiv>
  <refsection>
    <title>Arguments</title>
    <variablelist>
      <varlistentry>
        <term>x</term>
        <listitem>
          <para>
            Add here the input/output argument description.
          </para>
        </listitem>
      </varlistentry>
      <varlistentry>
        <term>y</term>
        <listitem>
          <para>
            Add here the input/output argument description.
          </para>
        </listitem>
      </varlistentry>
    </variablelist>
  </refsection>
  <refsection>
    <title>Description</title>
    <para>
          Add here a paragraph of the function description.
          Other paragraph can be added
    </para>
    <para>With a latex expression
          <latex>
            \begin{eqnarray}
    f(x,a,r) = \frac{1}{r^{-a}\Gamma(a)} \int_0^x t^{a-1} \exp\left(-rt
\right) dt
    \end{eqnarray}
    </latex>
    </para>
  </refsection>
  <refsection>
    <title>More information</title>
    <note><para>A note about foo</para></note>
    <caution><para>A caution about foo</para></caution>
    <warning><para>A warning about foo</para></warning>
    <important><para>An important about foo</para></important>
    <tip><para>A tip about foo</para></tip>
```

```
    </refsection>
    <refsection>
      <title>Examples</title>
      <programlisting role="example"><![CDATA[
          Add here scilab instructions and comments
      ]]></programlisting>
    </refsection>
    <refsection>
      <title>See Also</title>
      <simplelist type="inline">
        <member>
          <link linkend="add a reference name" >add a reference</link>
        </member>
        <member>
          <link linkend="add a reference name">add a reference</link>
        </member>
      </simplelist>
    </refsection>
    <refsection>
      <title>Authors</title>
      <simplelist type="vert">
        <member>add the author name and author reference</member>
        <member>add another author name and it's reference</member>
      </simplelist>
    </refsection>
    <refsection>
       <title>Bibliography</title>
         <para>
           Add here the function bibliography
         </para>
       </refsection>
      <refsection>
         <title>History</title>
        <revhistory>
          <revision>
            <revnumber>X.Y</revnumber>
            <revdescription>Function foo added</revdescription>
          </revision>
        </revhistory>
      </refsection>
    <refsection>
       <title>Used Functions</title>
         <para>
           Add here the Scilab, C,... used code references
         </para>
       </refsection>
  </refentry>
```

18

Example : Programming a Sudoku Game

In order to put to use the different concepts we covered in this part, we are going to program a game of sudoku. To do so, let's start with the example in Section 13.4, *Creating a sudoku*.

Figure 18.1 : *Sudoku example*

$$S = \begin{array}{|c|c|c||c|c|c||c|c|c|}\hline 5 & 3 & & & 8 & & & 2 & \\ \hline 8 & & & & 4 & 2 & & & \\ \hline & & 1 & 3 & & 6 & & 8 & \\ \hline\hline 6 & 5 & 3 & & & & 1 & & 2 \\ \hline 2 & 1 & 4 & 6 & & 3 & 5 & 7 & 8 \\ \hline 9 & & 8 & & & & 3 & 6 & 4 \\ \hline\hline & 6 & & 5 & & 1 & 8 & & \\ \hline & & & 4 & 6 & & & & 5 \\ \hline & 4 & & & 3 & & & 1 & 6 \\ \hline \end{array}$$

We represent a soduku with a matrix `S` with 9 rows and 9 columns, where each cell can contain:

- a value from 1 to 9 when the entry is filled
- the value 0 if the entry is empty

For example, for the sudoku in Figure 18.1:

```
S =[  5  3  0  0  8  0  0  2  0;
      8  0  0  0  4  2  0  0  0;
      0  0  1  3  0  6  0  8  0;
      6  5  3  0  0  0  1  0  2;
      2  1  4  6  0  3  5  7  8;
      9  0  8  0  0  0  3  6  4;
```

```
0  6  0  5  0  1  8  0  0;
0  0  0  4  6  0  0  0  5;
0  4  0  0  3  0  0  1  6];
```

Filling the empty elements must obey the following three rules:

- Each value from 1 to 9 appears once per column.
- Each value from 1 to 9 appears once per row.
- Each value from 1 to 9 appears once per region.

The game is solved once all the entries are filled by following these three rules (there are no more `0`s in the sudoku).

18.1. Functional programming

To simplify the program, we start by breaking it up into easier subproblems which can be solved by creating a function. This approach, often called functional programming, is very easy to implement in Scilab. This method becomes particularly interesting when a complicated operation reappears several times inside a problem. For example, in the case of a sudoku, a quick analysis reveals the presence of several recurring calculations:

Compute the region number `r` associated to an entry `(i,j)`

Let's create a function `pos2region` by copying the formulas used to construct the sudoku which yield the value of `r` as a function of `(i,j)`. All we need to do is put these formulas in the body of a function which takes the parameters `i,j` as inputs and outputs `r`. This yields:

```
function r=pos2region(i,j)
    // sudoku with 9 rows and 9 columns
    // r=number of the region associated to the element
    // (i,j)=entry position in the sudoku
    r=1+floor((j-1)/3)+floor((i-1)/3)*3
endfunction
```

We then perform a few tests to validate the function:

```
-->pos2region(1,1)  // region 1
 ans  =
    1.

-->pos2region(2,4)  // region 2
 ans  =
    2.

-->pos2region(3,7)  // region 3
 ans  =
    3.

-->pos2region(8,8)  // region 9
 ans  =
    9.
```

Compute the coordinates `(i1,j1)` and `(i2,j2)` of the cells delimiting the region number `r`

Here again, let's create a function which takes `r` as input and outputs the four values `i1,j1,i2,j2` corresponding to the upper-left and lower-right entries which delimit the region of interest. For this purpose we need to use the formulas:

- `i1= 1+ quotient (r -1)/3`
- `j1= 1+ remainder of (r -1)/3`

We then only need to add 2 to these values to get `i2` and `j2`. This yields:

```
function [i1,i2,j1,j2]=region2pos(r)
    // sudoku with 9 rows and 9 columns
    // r=region number
    // (i1:i2,j1:j2)=corresponding sudoku zone
    column=pmodulo(r-1,3)
    j1=1+column*3
    j2=3+column*3
    row=(r-1-column)/3
    i1=1+row*3
    i2=3+row*3
endfunction
```

We then perform a few tests to validate the function:

```
-->// region 1 ->  elements (1,1) and (3,3)

-->[i1,i2,j1,j2]=region2pos(1)
 j2  =
    3.
 j1  =
    1.
 i2  =
    3.
 i1  =
    1.

-->// region 4 ->  elements (1,7) and (3,7)

-->[i1,i2,j1,j2]=region2pos(4)
 j2  =
    3.
 j1  =
    1.
 i2  =
    6.
 i1  =
    4.

-->// region 8 ->  elements (7,4) and (9,6)

-->[i1,i2,j1,j2]=region2pos(8)
 j2  =
    6.
 j1  =
    4.
 i2  =
    9.
 i1  =
    7.
```

Compute the coordinates `(i,j)` of the element number `k`

If we number the sudoku elements column by column, we can reuse the formulas of chapter *Calculation Examples* where we constructed a matrix containing the position number `N(i,j)` of each element. If we define both the number of rows `n` and the number of columns `m` of the matrix as `9`, we can retrieve:

- the value of `i-1` from the remainder of `N(i,j)` in the division by 9 (computable with `pmodulo`)
- the value of `j-1` from the quotient of `N(i,j)` in the division by 9

We only need to put these commands inside the body of a function which takes `k` as input parameter and outputs `i,j`. This yields:

```
function  [i,j]=num2pos(k)
    // S = sudoku with 9 rows and 9 columns
    // k = element's position number
    // (i,j)=position of the entry in the sudoku
    i=1+pmodulo(k-1,9)
    j=1+(k-i)/9
endfunction
```

We then perform a few tests to validate the function:

```
-->// entry number 1 -> position (1,1)

-->[i,j]=num2pos(1)
 j  =
    1.
 i  =
    1.

-->// entry number 9 -> position (9,1)

-->[i,j]=num2pos(9)
 j  =
    1.
 i  =
    9.

-->// entry  number 18 -> position (9,2)

-->[i,j]=num2pos(18)
 j  =
    2.
 i  =
    9.

-->// entry number 73 -> position (1,9)

-->[i,j]=num2pos(73)
 j  =
    9.
 i  =
    1.

-->// entry number -> position (9,9)

-->[i,j]=num2pos(81)
 j  =
    9.
 i  =
    9.
```

Make a list `P` of possible values of a cell `(i,j)` in the sudoku `S`

This time again, we use the same commands and place them inside the body of a function which takes as input the data `S, i, j` and outputs a vector containing all the possible values for element `(i,j)`. This yields:

```
function P=possibility(i,j,S)
    // S= matrix with 9 rows and 9 columns = sudoku
    // (i,j)= cell of interest in sudoku S
    // P= list of possible values out of {1;2;...;9}
    // for element (i,j)
    S(i,j)=0   // we set element (i,j) to 0
    A=unique(S(i,:))  // values in row i
    A=A(find(A<>0))   // remove the value 0
    B=unique(S(:,j))  // values in column j
    B=B(find(B<>0))   // remove the value 0
    // values already used in the region
    r=pos2region(i,j)
    [i1,i2,j1,j2]=region2pos(r)
    C=unique(S(i1:i2,j1:j2))
    C=C(find(C<>0))   // remove the value 0
    // union of all values already used
    E=union(A,union(B,C))
    // remaining options
    P=[1:9],P(E)=[]
endfunction
```

We then perform a few tests to validate the function:

```
-->S=[ 5 3 0 0 8 0 0 2 0;
-->    8 0 0 0 4 2 0 0 0;
-->    0 0 1 3 0 6 0 8 0;
-->    6 5 3 0 0 0 1 0 2;
-->    2 1 4 6 0 3 5 7 8;
-->    9 0 8 0 0 0 3 6 4;
-->    0 6 0 5 0 1 8 0 0;
-->    0 0 0 4 6 0 0 0 5;
-->    0 4 0 0 3 0 0 1 6];

-->//  possible values of cell (1,3)  - >  6,7,9

-->P=possibility(1,3,S)
 P  =
    6.    7.    9.
```

Check if the value in cell `(i,j)` respects the rules for filling

We only need to test if the value of element `(i,j)` appears in the corresponding row, column or region. We can also check if its value indeed falls between 1 and 9, which yields:

```
function bool=test_cell(i,j,S)
   // S= sudoku
   // (i,j)= cell to check in sudoku S
   // bool= True if S(i,j) is consistent with the rest of the
 sudoku,
   // False otherwise
   value=S(i,j) // retrieve the value of S(i,j)
   S(i,j)=0      // set element (i,j) to 0
   row=S(i,:)
   column=S(:,j)
   r=pos2region(i,j)
   [i1,i2,j1,j2]=region2pos(r)
   region=S(i1:i2,j1:j2)
   bool=and(row<>value)&and(column<>value)&and(region<>value)
&(value<=9)&(value>0)
endfunction
```

Let's perform a few tests to validate the function:

```
-->S=[ 5 3 0 0 8 0 0 2 0;
-->    8 0 0 0 4 2 0 0 0;
-->    0 0 1 3 0 6 0 8 0;
-->    6 5 3 0 0 0 1 0 2;
-->    2 1 4 6 0 3 5 7 8;
-->    9 0 8 0 0 0 3 6 4;
-->    0 6 0 5 0 1 8 0 0;
-->    0 0 0 4 6 0 0 0 5;
-->    0 4 0 0 3 0 0 1 6];

-->//  possible value of cell (1,3)  - >  6,7,9

-->// test_cell answers true

-->S(1,3)=7; test_cell(1,3,S)
 ans  =
  T

-->// test_cell answers false

-->S(1,3)=4; test_cell(1,3,S)
 ans  =
  F
```

In what follows, we are going to time after time call these four functions.

Tip › Breaking up a problem into simpler functions helps make it easier to elaborate complex programs. Specifically, in order to simplify the debugging of more complex functions, it is very important that you perform test as thoroughly as possible on these basic functions.

18.2. Solving a game of sudoku

Starting from the functions seen above, we want to create a program to play sudoku. The program will take as input a matrix representing a sudoku and will then need to let the user perform several tasks:

- Fill one or several empty sudoku cells.
- Undo the previous move.
- Display in the console the current state of the sudoku.
- Display in the console the possible values of each cell.
- Save the game.

In order to practice, first try to write this program on your own. To do this, break the problem up into three functions:

- A main function `play_sudoku` which takes as argument a sudoku `S` and a list `L` which contains all the moves already played (in order to undo them) – this function will call itself throughout the game until it is over.
- A function `enter_values` which lets you enter sudoku values and takes as argument the sudoku `S` and the initial sudoku `S0` (to make sure we are not modifying elements of the starting sudoku) and outputs the modified sudoku – this function will need to check that all the suggested values are compatible with the rest of the sudoku.
- A function `see_possibility` which takes as argument the sudoku `S` and lets the user see the possible values for each of the sudoku's cell in the console.

Tip › Before starting, here is a little advice on the functions to use:

- *To store the different steps of the game (and potentially undo them) you can use a list **L** and store each new version of the sudoku inside it.*
- *Write the main function **`play_sudoku(S,L)`** following a recursive format. At each step, the function calls itself and passes as input a sudoku containing the latest modifications and a new list of the previous moves.*
- *Use a dialog box **`x_mdialog`** to enter the sudoku values.*
- *Use a dialog box **`x_choose`** to choose the task to perform.*
- *Use **`disp`** to display console messages or **`messagebox`** to display important messages for the user.*

- *To save the game, save the variables* **S** *and* **L** *inside a* ***.sod** *file with the command* **save**. *The user will then only need to reload these variables with the command* **load** *and enter the command* **play_sudoku(S,L)** *to resume the game.*

The easiest function to write is the function `see_possibility`. Create, then display (with `disp`) a strings matrix of the same size as the sudoku matrix and fill each element with:

- the list of possible values for this cell, if the cell is empty – you can use the function `possibility` to get this list then convert it to a string with `string` and `strcat`
- the cell's value if it has one assigned

Here is a possible version of this function:

```
function see_possibility(S)
    // S= sudoku  with 9 rows and 9 columns
    S_pos=[]
    for i=1:9
        row=[]
        for j=1:9
            if S(i,j)==0 then P=possibility(i,j,S)
            else P=S(i,j)
            end
            row=[row strcat(string(P),',')]
        end
        S_pos=[S_pos;row]
    end
    disp('here is the list of possible values for each cell :')
    disp(S_pos)
endfunction
```

Since it's not easy to create strings of a given size, create a matrix by concatenation, row by row. You will get the following display:

```
-->    8 0 0 0 4 2 0 0 0;
-->    0 0 1 3 0 6 0 8 0;
-->    6 5 3 0 0 0 1 0 2;
-->    2 1 4 6 0 3 5 7 8;
-->    9 0 8 0 0 0 3 6 4;
-->    0 6 0 5 0 1 8 0 0;
-->    0 0 0 4 6 0 0 0 5;
-->    0 4 0 0 3 0 0 1 6];

-->see_possibility(S)

here is the list of possible values for each cell :

!5     3      6,7,9  1,7,9  8      7,9    4,6,7,9 2     1,7,9  !
!                                                            !
!8     7,9    6,7,9  1,7,9  4      2      6,7,9  3,5,9 1,3,7,9!
```

```
!                                                                        !
!4,7    2,7,9    1        3        5,7,9    6        4,7,9    8      7,9    !
!                                                                        !
!6      5        3        7,8,9    7,9      4,7,8,9  1        9      2      !
!                                                                        !
!2      1        4        6        9        3        5        7      8      !
!                                                                        !
!9      7        8        1,2,7    1,2,5,7  5,7      3        6      4      !
!                                                                        !
!3,7    6        2,7,9    5        2,7,9    1        8        3,4,9  3,7,9  !
!                                                                        !
!1,3,7  2,7,8,9  2,7,9    4        6        7,8,9    2,7,9    3,9    5      !
!                                                                        !
!7      4        2,5,7,9  2,7,8,9  3        7,8,9    2,7,9    1      6      !
```

Let's now move on to the function `enter_values`. First use `x_mdialog` to enter the sudoku's values. We will then need to check that the values' compability:

- with `find`, you can easily retrieve the positions of the modified elements and check that they are empty cells in the starting sudoku
- with `test_cell` you can check that the values entered are compatible with the sudoku's remaining values

Here is a possible version of this function:

```
function S=enter_values(S,S0)
    // S  = sudoku's current state
    // S0 = initial sudoku
    // call from the variable editor to modify the sudoku
    text=[ 'enter new values for the sudoku ...
    (only modify cells that contain a 0) :';
    '';
    strcat(string(S0),' ','c')]
    S1=evstr(x_mdialog(text,string(1:9),string(1:9),string(S)))
    // check that the modification is valid
    [I,J]=find((S1<>S)&(S0<>0))
    if I<>[] then  // we modified cells that were not allowed
        messagebox('you have modified the sudoku''s initial given
 values!',...
        'error','modal')
    else  // check the validity of the new sudoku
        [I,J]=find((S1<>S))
        bool=%t
        while bool & (I<>[])
            // for each change
            i=I(1),I(1)=[] // get the row
            j=J(1),J(1)=[] // get the column
            bool=test_cell(i,j,S1)  // check
        end
        if bool then // verify the change
             S=S1
         else // if it is incompatible, undo modifications
```

```
                messagebox('the value '+string(S1(i,j))+' of cell
    ('+string(i)+','+string(j)+') is not permitted!','error','modal')
            end
        end
    endfunction
```

You can try out the function with the command `enter_values(S,S)`. You will get an interface prompt similar to the one in Figure 18.2.

Figure 18.2 : Interface to enter sudoku values

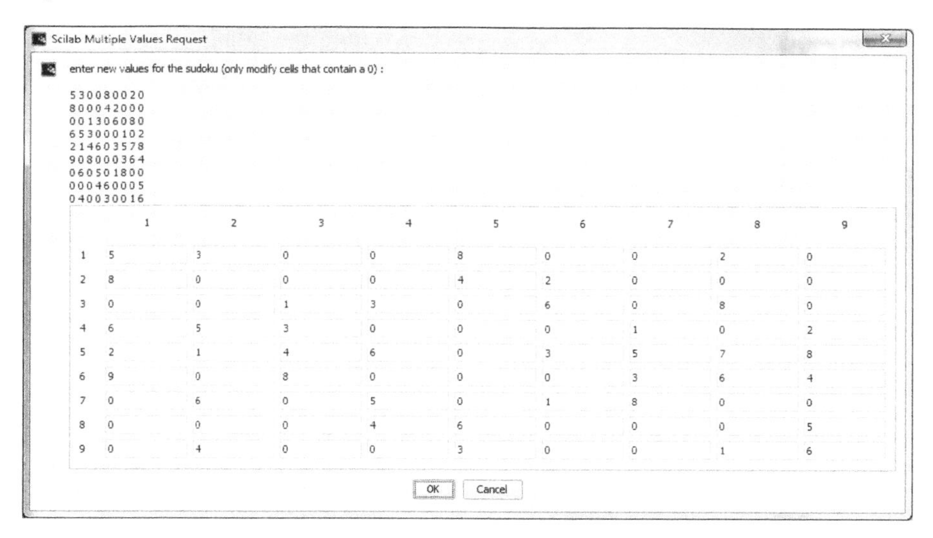

If there are any errors while values are entered, a `messagebox` dialog box gets displayed to explain the problem. For example, if you enter the value 3 for cell (1,3), you will get the message in Figure 18.3.

Figure 18.3 : Message indicating a user input error

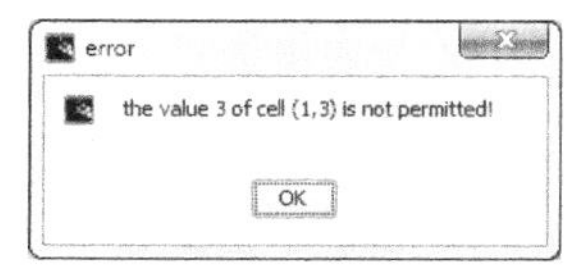

Similarly, if you try to change the value of a cell that did not initially contain a zero, the function shows the same message as in Figure 18.4 and outputs the unchanged sudoku:

```
-->S=[ 5 3 0 0 8 0 0 2 0;
-->    8 0 0 0 4 2 0 0 0;
-->    0 0 1 3 0 6 0 8 0;
-->    6 5 3 0 0 0 1 0 2;
-->    2 1 4 6 0 3 5 7 8;
-->    9 0 8 0 0 0 3 6 4;
-->    0 6 0 5 0 1 8 0 0;
-->    0 0 0 4 6 0 0 0 5;
-->    0 4 0 0 3 0 0 1 6];

-->S1=enter_values(S,S)
 S1  =

    5.    3.    0.    0.    8.    0.    0.    2.    0.
    8.    0.    0.    0.    4.    2.    0.    0.    0.
    0.    0.    1.    3.    0.    6.    0.    8.    0.
    6.    5.    3.    0.    0.    0.    1.    0.    2.
    2.    1.    4.    6.    0.    3.    5.    7.    8.
    9.    0.    8.    0.    0.    0.    3.    6.    4.
    0.    6.    0.    5.    0.    1.    8.    0.    0.
    0.    0.    0.    4.    6.    0.    0.    0.    5.
    0.    4.    0.    0.    3.    0.    0.    1.    6.
```

Figure 18.4 : *Message to warn of a change in the sudoku's initial values*

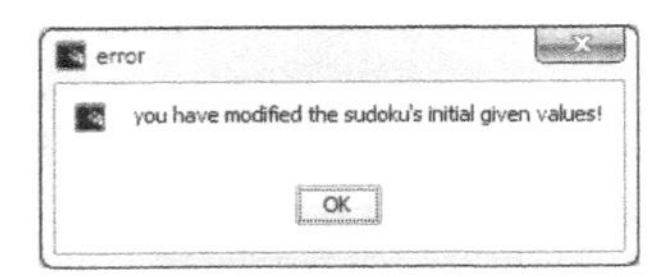

We use the `'modal'` argument for the `messagebox` command which prevents the game from resuming as long as the user hasn't clicked on `'OK'`. Finally, the function left to write is the main function `play_sudoku`. This function is naturally recursive since it calls itself with different arguments depending on the move played (or undone). The list of previous moves can be seen as an optional argument which is not provided at the start of the game.

Tip › Here are a few tips to write this function:

- *At the beginning, the list of previous moves can be stored inside the variable* **`varargin`** *(or create it if it is empty).*
- *To choose the type of move to perform, use the command* `x_choose`.
- *The number returned by* `x_choose` *can be processed using* **`select`**.
- *For each of the* **`select`** *cases, call* **`play_sudoku`** *with new arguments.*

Here is a possible version of this function:

```
function play_sudoku(S,varargin)
    // create or get the list of previous moves
```

```
    if length(varargin)>0 then L=varargin(1)
    // at the beginning, store S inside the list of previous moves
    else L=list(S)
    end
    // choose the next move
    item=['choose a value ';                        //  num=1
    'undo the previous move';                       //  num=2
    'display the options for each cell';            //  num=3
    'display the sudoku';                           //  num=4
    'save the game';]                               //  num=5
    text=['double-click your choice']
    num=x_choose(item,text,'quit')
    // process the choice
    select num
    case 1 then
        S1=enter_values(S,L($))
        if and(S1==S)  then   // no changes!
        // play again without modifying to list of moves
            play_sudoku(S,L)
        // all the cells are filled correctly
        elseif and(S1<>0) then
            disp('you won!')      // you won
        else                    // keep playing
            L($+1)=S    // add the previous move to the list
            play_sudoku(S1,L) // play again
        end
    case 2 then                 // undo the previous move
        if size(L)>1 then       // we can undo the previous move
            S0=L($)       // retrieve the previous state of the sudoku
            L($)=null()         // then delete it from the list
            disp('last choice canceled')
        else S0=S               // S  is the initial sudoku
            L=list()            // list of previous moves is empty
            disp('you are back to the sudoku''s initial state!')
        end
        play_sudoku(S0,L)       // play again
    case 3 then see_possibility(S)   // display options
        play_sudoku(S,L)             // play again
    case 4 then              // display the sudoku in the console
        K=string(S)          // convert S to a strings
        K(find(S==0))='_'  // find empty cells
        disp(K,'current sudoku state :')
        play_sudoku(S,L)  // then play again
    case 5 then // save the game
        path=uiputfile('*.sod',pwd(),'choose a file to save the game:')
        if path<>'' then    //  a path is entered correctly
            save('''+path+''','S','L') // save S and L
            disp('game saved')
            else disp('game not saved')
        end
    else // abort the game without saving
        disp('bye!')
    end
endfunction
```

The main menu should resemble the following figure.

Figure 18.5 : Main menu of sudoku game

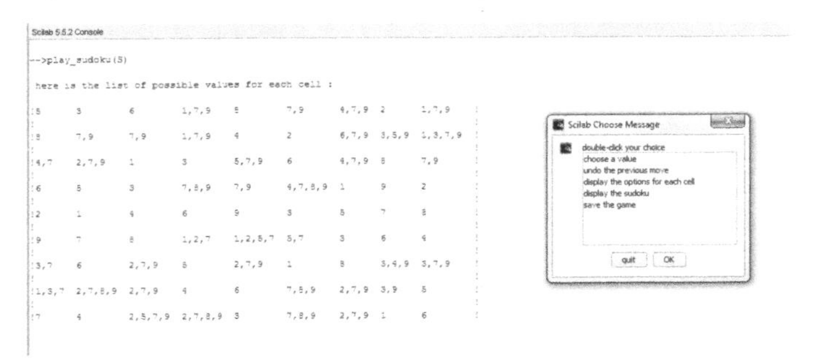

To keep playing a game you saved, you'll need to:

- Use the `load` command to load the file containing the old game.
- Relaunch the game with `play_sudoku(S,L)` to keep track of the list of earlier moves.

Here's an example of saving a game and reloading it after four moves were played:

```
-->play_sudoku(S)

 here is the list of possible values for each cell:

!5     3     6     1,7,9   8       7,9     4,7,9  2      1,7,9  !
!                                                                !
!8     7     9     1,9     4       2       6,9    3,5,9  1,3,9  !
!                                                                !
!4     2,9   1     3       5,7,9   6       4,7,9  8      7,9    !
!                                                                !
!6     5     3     7,8,9   7,9     4,7,8,9 1      9      2      !
!                                                                !
!2     1     4     6       9       3       5      7      8      !
!                                                                !
!9           8     1,2,7   1,2,5,7 5,7     3      6      4      !
!                                                                !
!3,7   6     2,7,9 5       2,7,9   1       8      3,4,9  3,7,9  !
!                                                                !
!1,3,7 2,8,9 2,7,9 4       6       7,8,9   2,7,9  3,9    5      !
!                                                                !
!7     4     2,5,7,9 2,7,8,9 3     7,8,9   2,7,9  1      6      !

 game saved

-->load('sudoku.sod')  //  load the saved game again

-->S
 S  =

    5.    3.    6.    1.    8.    0.    0.    2.    0.
```

```
    8.    7.    0.    0.    4.    2.    0.    0.    0.
    0.    0.    1.    3.    0.    6.    0.    8.    0.
    6.    5.    3.    0.    0.    0.    1.    0.    2.
    2.    1.    4.    6.    0.    3.    5.    7.    8.
    9.    0.    8.    0.    0.    0.    3.    6.    4.
    0.    6.    0.    5.    0.    1.    8.    0.    0.
    0.    0.    0.    4.    6.    0.    0.    0.    5.
    0.    4.    0.    0.    3.    0.    0.    1.    6.

-->length(L)
 ans  =

    4.

-->typeof(L)
 ans  =

 list

-->L
 L  =

       L(1)

    5.    3.    0.    0.    8.    0.    0.    2.    0.
    8.    0.    0.    0.    4.    2.    0.    0.    0.
    0.    0.    1.    3.    0.    6.    0.    8.    0.
    6.    5.    3.    0.    0.    0.    1.    0.    2.
    2.    1.    4.    6.    0.    3.    5.    7.    8.
    9.    0.    8.    0.    0.    0.    3.    6.    4.
    0.    6.    0.    5.    0.    1.    8.    0.    0.
    0.    0.    0.    4.    6.    0.    0.    0.    5.
    0.    4.    0.    0.    3.    0.    0.    1.    6.

       L(2)

    5.    3.    0.    0.    8.    0.    0.    2.    0.
    8.    0.    0.    0.    4.    2.    0.    0.    0.
    0.    0.    1.    3.    0.    6.    0.    8.    0.
    6.    5.    3.    0.    0.    0.    1.    0.    2.
    2.    1.    4.    6.    0.    3.    5.    7.    8.
    9.    0.    8.    0.    0.    0.    3.    6.    4.
    0.    6.    0.    5.    0.    1.    8.    0.    0.
    0.    0.    0.    4.    6.    0.    0.    0.    5.
    0.    4.    0.    0.    3.    0.    0.    1.    6.

       L(3)

    5.    3.    6.    0.    8.    0.    0.    2.    0.
    8.    0.    0.    0.    4.    2.    0.    0.    0.
    0.    0.    1.    3.    0.    6.    0.    8.    0.
    6.    5.    3.    0.    0.    0.    1.    0.    2.
    2.    1.    4.    6.    0.    3.    5.    7.    8.
    9.    0.    8.    0.    0.    0.    3.    6.    4.
    0.    6.    0.    5.    0.    1.    8.    0.    0.
    0.    0.    0.    4.    6.    0.    0.    0.    5.
```

```
    0.    4.    0.    0.    3.    0.    0.    1.    6.

         L(4)

    5.    3.    6.    0.    8.    0.    0.    2.    0.
    8.    7.    0.    0.    4.    2.    0.    0.    0.
    0.    0.    1.    3.    0.    6.    0.    8.    0.
    6.    5.    3.    0.    0.    0.    1.    0.    2.
    2.    1.    4.    6.    0.    3.    5.    7.    8.
    9.    0.    8.    0.    0.    0.    3.    6.    4.
    0.    6.    0.    5.    0.    1.    8.    0.    0.
    0.    0.    0.    4.    6.    0.    0.    0.    5.
    0.    4.    0.    0.    3.    0.    0.    1.    6.

-->play_sudoku(S,L)  // relaunch the game
```

18.3. Solving a sudoku automatically

Many complicated problems can be resolved by testing all possible combinations. If the number of possibilities that need to be tested is acceptable, this kind of solution can be programmed with in Scilab. The issue then is figuring out in which order to test all the combinations. In the sudoku's case, here is a fairly simple algorithm that lets you test all the possible combinations starting from an initial table.

Figure 18.6 : Greedy algorithm to solve a sudoku

<u>**function**</u> $[S]$ = solve_sudoku(S)
  Initialization :
  $Free$ = boolean table : True if cell (i,j) is empty, False otherwise
  $k = 1$ (current cell number)
  $upordown = 1$ (sudoku displacement direction)
  Processing :
  <u>**while**</u> $k \leq 9^2$ <u>**do**</u>
    (i,j) = position of cell number k
    <u>**if**</u> $Libre(i,j)$ <u>**then**</u> check the next value for this cell : $S(i,j) = S(i,j) + 1$
      <u>**if**</u> the value $S(i,j)$ is consistent with the rest of the sudoku
        <u>**then**</u> move on to the next cell : $upordown = 1$
        <u>**else if**</u> $S(i,j) > 9$ <u>**then**</u> we have checked all possible values without success
          free up the cell : $S(i,j) = 0$
          go back a step $upordown = -1$
        <u>**else**</u> stay on this cell : $upordown = 0$
        <u>**end**</u>
      <u>**end**</u>
    <u>**end**</u>
    move on to cell $k = k + upordown$
  <u>**end do**</u>

Now it's your turn to translate this algorithm to Scilab's language! Each line of this algorithm roughly corresponds to a Scilab command or a function already created in the previous examples.

Caution › *When you translate an algorithm described in pseudocode or in another programming language, make sure you follow the Scilab syntax. In this example, pay attention to the conditional expressions* **`if-then-else`** *and remember that in Scilab, you can omit the* **`then`**, *however when it is used, it must always be on the same line as the* **`if`**.

Here's a possible implementation of this algorithm:

```
function [S]=solve_sudoku(S)
    Free=(S==0)  // initial sudoku coefficient positions
    k=1            // cell number of interest
    upordown=1     // to manage the progression forwards or backwards
    while k<=9^2  // as long as we haven't looked at all cells
        [i,j]=num2pos(k)       // position of cell number k
        if Free(i,j) then     // if cell (i,j) is empty
            S(i,j)=S(i,j)+1   // increment the value
            if test_cell(i,j,S) then // if it passes the test
                upordown=1         // move on to the next cell
     // we have checked all possible values:
            elseif S(i,j)>9 then
                S(i,j)=0          // free up the cell
                upordown=-1       // go back a step
            else  upordown=0      // otherwise stay on this cell
            end
        end
        k=k+upordown  // move on to the next cell
    end
endfunction
```

This very simple program is capable of solving relatively simple sudokus such as this one in just a few seconds:

```
-->S=[ 5 3 0 0 8 0 0 2 0;
-->    8 0 0 0 4 2 0 0 0;
-->    0 0 1 3 0 6 0 8 0;
-->    6 5 3 0 0 0 1 0 2;
-->    2 1 4 6 0 3 5 7 8;
-->    9 0 8 0 0 0 3 6 4;
-->    0 6 0 5 0 1 8 0 0;
-->    0 0 0 4 6 0 0 0 5;
-->    0 4 0 0 3 0 0 1 6];

-->solve_sudoku(S)
 ans  =
    5.    3.    6.    7.    8.    9.    4.    2.    1.
    8.    9.    7.    1.    4.    2.    6.    5.    3.
    4.    2.    1.    3.    5.    6.    7.    8.    9.
    6.    5.    3.    8.    7.    4.    1.    9.    2.
    2.    1.    4.    6.    9.    3.    5.    7.    8.
    9.    7.    8.    2.    1.    5.    3.    6.    4.
    3.    6.    9.    5.    2.    1.    8.    4.    7.
    1.    8.    2.    4.    6.    7.    9.    3.    5.
    7.    4.    5.    9.    3.    8.    2.    1.    6.
```

It also manages to solve sudokus of medium difficulty level in a few minutes but is incapable of solving difficult sudokus in a reasonable amount of time. It is of course possible to write a Scilab program to solve sudokus whose level is hard or evil, however you will need a more complex algorithm and a more optimized implementation.

Creating Plots

Ever since its creation, Scilab has possessed numerous graphics capabilities and their management has largely evolved over the different software versions. The current graphics mode emerged in Scilab's version 4 and definitely replaced the previous graphics mode in the transition to version 5. This change has led to the removal of a lot of commands that were used by the old graphics mode and has greatly improved Scilab's graphics performance.

Tip › *The main figures in this part are available in color in the online gallery [http://d-booker.jo.my/scilab1-images-en].*

19

Graphics Entities and Windows

In the chapter Section 3.3, *Other windows*, we mentioned the presence of graphics functionalities in Scilab when talking about the graphics window Figure 19.1. In this chapter we will study in more detail how the graphics shown in this window are displayed. We will cover how to interact with this window by using other Scilab objects defined in the previous chapters.

***Figure 19.1** : Scilab's graphics window*

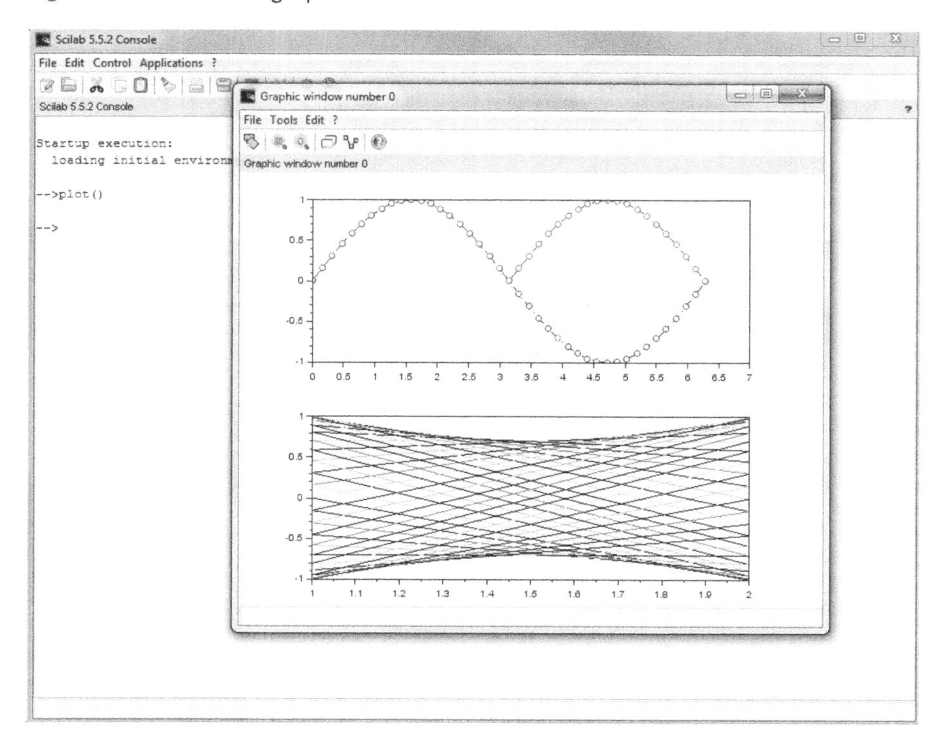

19.1. Variables of type *handle*

Objects that can get displayed in a window are stored inside variables that belong to a particular type: the type *handle*. This type of variable lets you store the properties of a graphics object and link them to other graphics entities that compose a figure (in effect, a *handle* is a pointer). Handles are hierarchically organized, as shown in Figure 19.2, where each entity has:

- a list of properties
- one or no parent
- several, one or no descendants, also called children

Figure 19.2 : *Hierarchic model of graphics entities*

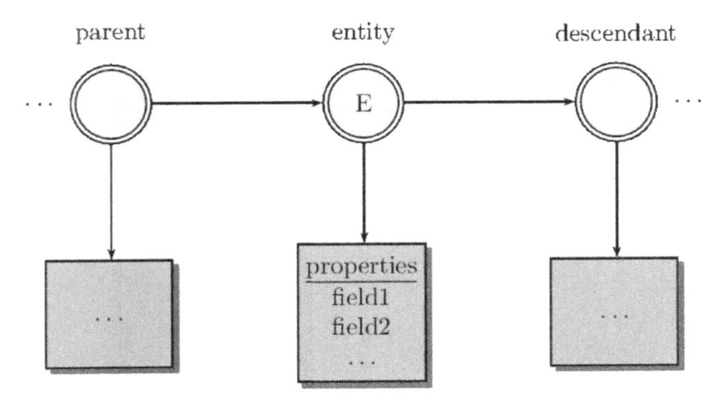

Tip › *To sort through all this data, it's useful to display the hierarchical structure of a figure's handles as a tree. Scilab's* **graphics entities editor** *was purposely made to simplify this task (see figure* **below** *or the video in* **Figure 3.8***). For Windows and Linux users, you can launch it from graphics window's EDIT menu or with the command* **ged** *from the console. Unfortunately, this editor isn't available yet for Mac OS users.*

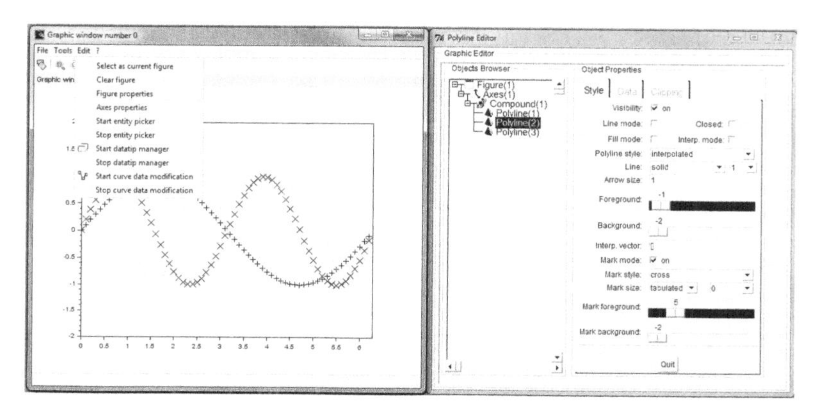

To interact with a figure and its entities, you need to access the associated handles and modify them by using:

- the `get` command to retrieve a handle
- the `set` command to modify a handle

Tip › In practice, the **`get`** *and* **`set`** *commands are often replaced with* **specific functions** *that eliminate the need to enter arguments and greatly simplifies its use.*

Concretely, the content of a graphics window is a complex tree with handles of different types. For each window, there is:

- a Figure handle which defines the graphics window and its basic properties (name, size, position, etc.). It does not have any parent but has one or several descendants belonging to the *axes* type.
- one or several Axes handles which indicates in which context the graphics objects that make up each figure are drawn (for example the position and visibility of coordinate axes, the scale, etc.). They have a handle parent of type *figure* and can have numerous children of different types.
- Compound handles, which have a handle of type *axes* and one or several children of a different type. It is used to group together several simple graphics entities to create a more complex entity. For example, a figure with a set of curves is a *compound* type handle which contains handles of type *polyline* that represent each of the curves.

All other graphics entities have a parent of type *axes* or *compound* and do not have any descendants. Depending on their type, they describe the different objects that can be drawn in Scilab (*polyline* to draw lines, *text* to write text, etc.).

To access these handles, you will primarily use the commands:

- `gcf` or `get("current_figure")` to retrieve the current figure's Figure handle
- `gca` or `get("current_axes")` to retrieve the Axes handle of the current axes
- `gce` or `get("current_entities")` to retrieve the handle of the last graphics entity created (of type *compound* or other)

```
-->plot()    // Scilab graphics command

-->gcf()     // handle of corresponding figure
 ans  =
Handle of type "Figure" with properties:
========================================
figure_position = [200,200]
figure_size = [626,586]
axes_size = [610,460]
auto_resize = "on"
viewport = [0,0]
figure_name = "Graphic window number %d"
figure_id = 0
info_message = ""
color_map = matrix 37x3
pixel_drawing_mode = "copy"
anti_aliasing = "off"
immediate_drawing = "on"
background =  -2
visible = "on"
rotation_style = "unary"
event_handler = ""
event_handler_enable = "off"
user_data = []
resizefcn = ""
closerequestfcn = ""
resize = "on"
toolbar = "figure"
toolbar_visible = "on"
menubar = "figure"
menubar_visible = "on"
infobar_visible = "on"
dockable = "on"
layout = "none"
layout_options = "OptNoLayout"
default_axes = "on"
icon = ""
tag = ""
```

Caution › *The length of the content for each handle displayed, as shown in the example* **above**, *gives a good idea of the large amount of information stored inside variables of type* handle! *To prevent it from displaying this data, you can use:*

- *the semicolon* ; *to block the display of results (see* **Section 2.2, Using the console***)*

- *the* **lines(k)** *command which asks you whether you want to keep displaying after* **k** *lines or if you want to return to the prompt*

```
-->lines(10)    // limits display to 10 lines

-->plot()       // Scilab graphics command

-->gcf()        // handle of corresponding figure
 ans  =
Handle of type "Figure" with properties:
========================================
figure_position = [200,200]
figure_size = [626,586]
axes_size = [610,460]
auto_resize = "on"
viewport = [0,0]
```

When Scilab pauses the display, you can just type **n** *(no) in the console to stop or* **y** *(yes) to resume. If you prefer Scilab to display all the results without asking for your input, use the* **lines(0)** *command.*

Once you close a graphics window, its associated handle disappears and it can no longer be accessed. However, if you try, you will get an error (999) with warning *get: The handle is not or no more valid.*

```
-->plot();  // opening the graphics window

-->F=gcf()  // handle of type Figure
 F  =
Handle of type "Figure" with properties:
========================================
figure_position = [200,200]
figure_size = [626,586]
axes_size = [610,460]
auto_resize = "on"
viewport = [0,0]
-->// close the graphics window

-->delete(F);     //  to delete

-->F        //we no longer have access to F
 ans  =

    !--error 999

 get: The handle is not or no more valid.
```

Tip › *Rather than repeatedly closing the graphics window, particularly if you it docked to the console (see* **Figure 2.2** *in chapter* **The Console***), you can erase its content with the command* `clf` .

Finally, here is a list of commands you will often need to use:

To create new graphics entities or change the default parameters of the main graphics entities

- `figure` creates a new empty graphics window.
- `newaxis` creates a new Axes handle inside the current graphics window.
- `scf` or `set("current_figure")` defines a new default graphics window.
- `sca` or `set("current_axes")` defines a new default Axes handle.

To modify the handles' hierarchy

- `copy` lets you copy a handle.
- `delete` removes a handle. If it refers to a Figure handle, the corresponding graphics window gets closed.
- `glue` combines several handles into a handle of type *compound* and `unglue` breaks up the handles that make up a *compound* handle.
- `swap_handle` exchanges two handles' order in the graphics hierarchy and `relocate_handle` modifies a handle's parent.
- `move` moves a graphics entity within a handle of type *axes*.

Tip › *The* **`twinkle`** *command makes the graphics entity flash for a few seconds. This can be very useful to spot a particular graphics entity inside a complicated figure.*

19.2. First handle examples

To get used to dealing with handles, let's examine the figures generated by the graphics commands `surf()`, `plot2d()` and `plot()`.

Note › *As you can see in the examples, launching the graphics command automatically opens a graphics window. The default figure shown depends on the command. The most common commands are:*

- *`plot()` to draw a curve in a plane.*
- *`surf()` to draw a surface.*
- *`param3d()` to draw a curve in space.*

Example 19.1 : *Figure generated by the `surf()` command*

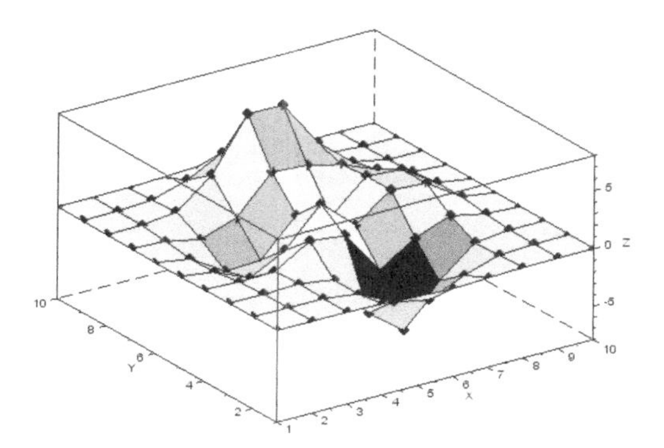

You can retrieve handles associated to a figure in the Scilab console by using the `gcf()`, `gca()` and `gce()` commands.

```
-->F=gcf()  // figure
 F  =
Handle of type "Figure" with properties:
========================================
figure_position = [200,200]
figure_size = [626,586]
axes_size = [610,460]
auto_resize = "on"
viewport = [0,0]
-->A=gca()  // axes
 A  =
Handle of type "Axes" with properties:
======================================
children: "Fac3d"

visible = "on"
axes_visible = ["on","on","on"]
axes_reverse = ["off","off","off"]
-->E=gce()  // handle of type Fac3D
 E  =
Handle of type "Fac3d" with properties:
=======================================
children: []
visible = "on"
surface_mode = "on"
foreground = 65
thickness = 1
```

In this example, the figure is made up of only three handles that can be displayed as in Figure 19.3.

Figure 19.3 : *Handles that make up the figure generated by the* `surf()` *command*

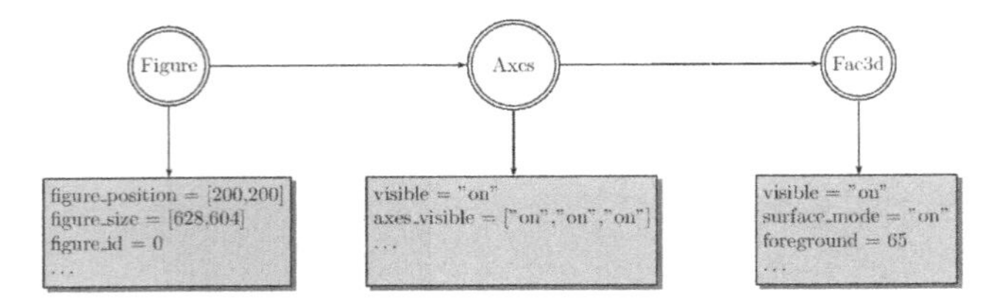

Example 19.2 : *Figure created by the* `plot2d()` *command*

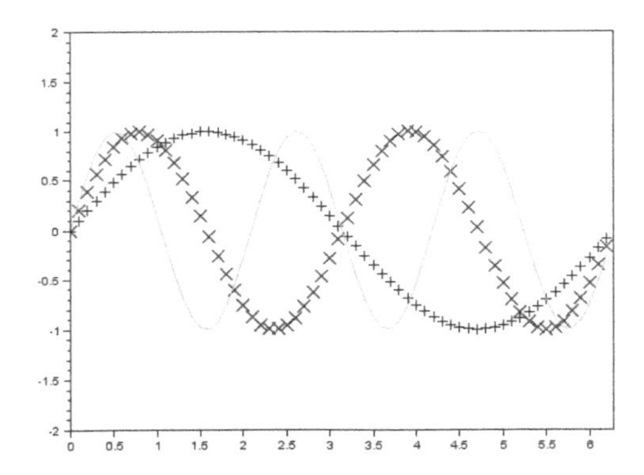

In this case, you get the following handles:

```
-->plot2d()

-->F=gcf()  // figure
 F  =
Handle of type "Figure" with properties:
========================================
figure_position = [200,200]
figure_size = [626,586]
axes_size = [610,460]
auto_resize = "on"
viewport = [0,0]
-->A=gca()  // axes
 A  =
Handle of type "Axes" with properties:
======================================
children: "Compound"

visible = "on"
axes_visible = ["on","on","on"]
axes_reverse = ["off","off","off"]
-->E=gce()  // handle of type Fac3D
```

```
 E =
Handle of type "Compound" with properties:
==========================================
children: ["Polyline","Polyline","Polyline"]
visible = "on"
user_data = []
tag =
```

The figure is consequently made up of six handles:

- two handles: Figure and Axes
- one Compound handle which combines three Polyline handles that create the three curves displayed

We can portray this as shown in Figure 19.4.

Figure 19.4 : *Handles that make up the figure generated by the* `plot2d()` *command*

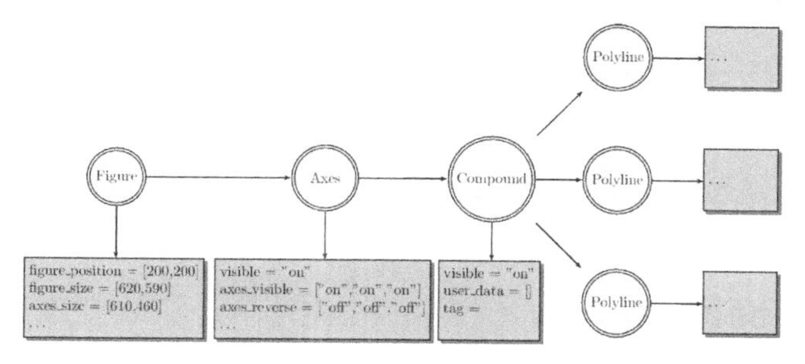

Example 19.3 : *Figure created by the* `plot()` *command*

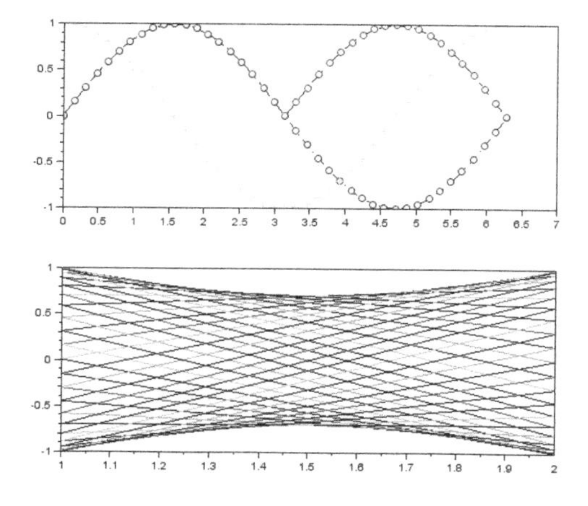

This time, the figure displays two independent plots which have two corresponding Axes handles. These two handles each point to one Compound handle which contains all the different Polyline handles that constitute all the curves drawn! Depicting all these handles then becomes fairly complex, as shown in Figure 19.5.

Figure 19.5 : *Handles that make up the figure generated by the `plot()` command*

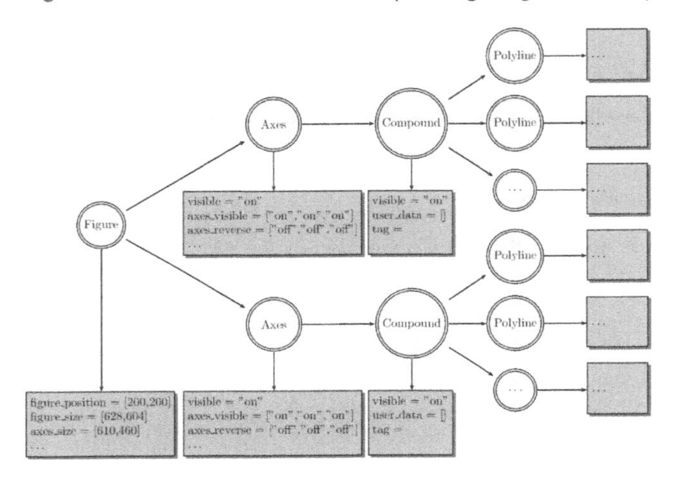

19.3. Handle Properties

To interact with a graphics object defined by a handle, you need to modify the properties that it is linked to. You can access the value of the property, called `prop`, inside the handle `H` or assign it a value `val` by using one of the following formats:

```
//retrieve a value
get(h,"prop")
h.prop
//modify a value
set(h,"prop",val)
h.prop=val
```

For example:

```
-->plot();F=gcf();

-->F.type            //  type of handle F= Figure
 ans  =
 Figure
```

```
-->F.visible        // on  <=> the figure is visible
 ans  =
 on

-->F.visible="off";// off <=> the figure becomes invisible

-->F.figure_id      // figure number 0
 ans  =
    0.

-->F.figure_id=1;  // in the graphics window, the number changes to 1
```

In this example, once you switch the `visible` property's value from `"on"` to `"off"`, the graphics entities of the figure of handle `F` are no longer displayed. Nevertheless, the figure hasn't been erased or deleted.

*Tip › The most convenient method of accessing and modifying handle properties consists in using a period "." rather than the commands **get** and **set**. This syntax gets around having to use quotation marks every time you write the **prop** string which signals the property you wish to access, such as in get("prop") or set(val, "prop").*

Let's look at a second example:

```
-->clf;plot()                    // Scilab graphics command

-->F=gcf();                      // handle of associated figure

-->A=F.children(1);              // Axes of the last plot

-->E=A.children                  // made up of 41 polylines
 E  =
Handle of type "Compound" with properties:
==========================================
children: matrix 41x1
visible = "on"
user_data = []
tag =

-->H=E.children(2);              // one of the 41 curves

-->H.thickness                   // curve thickness
 ans  =
    1.

-->H.thickness=10;               // modifying H.thickness

-->H.thickness                   // curve thickness is modified
 ans  =
    10.
```

This time, we modified the `thickness` property of the handle linked to one of the 41 curves of the second Axes generated by the `plot()`command. This immediately modifies the thickness of the curve drawn in the graphics window, as shown in Figure 19.6 (compare it to Example 19.3).

***Figure 19.6 :** Modifying the thickness of the curve in the figure generated by the* `plot()` *command*

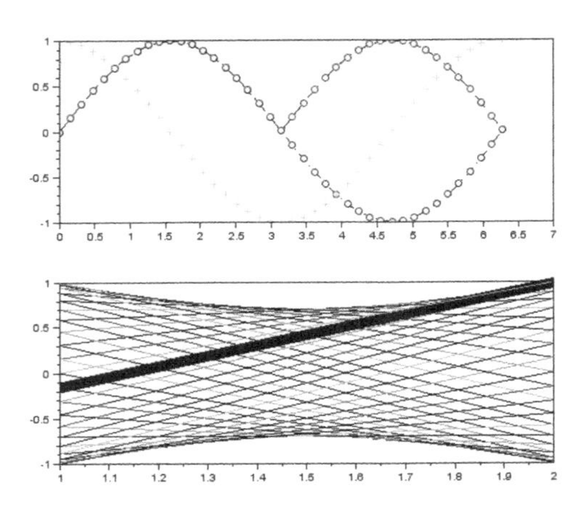

Each handle `H`'s property is a matrix that contains data of type *constant* or *string* (occasionally with just one data) or even `handle`. For example, if `H` is the figure handle generated by `plot()`, then:

- `H.type` is a string;
- while `H.figure_id` is of type *constant*;
- and `H.children` is a handle.

As you can do with all types of Scilab variables, you can construct matrices of handles. In addition, when a handle `H` has several children, the property `H.children` is a matrix of handles. The matrix elements are accessed with the use of parentheses `()` or with the use of a period "." to get to the result's properties (if it is a handle).

***Tip** › We can therefore use the "." recursively to retrieve a handle from one of its parents:*

```
-->clf;
-->plot();          // figure with two plots
-->F=gcf()          // handle Figure
 F  =
```

```
Handle of type "Figure" with properties:
========================================
figure_position = [200,200]
figure_size = [626,586]
axes_size = [610,460]
auto_resize = "on"
viewport = [0,0]
-->A=gca()         // handle Axes child of F
 A  =
Handle of type "Axes" with properties:
======================================
children: "Compound"

visible = "on"
axes_visible = ["on","on","on"]
axes_reverse = ["off","off","off"]
-->F.children(1)  // equal to A
 ans  =
Handle of type "Axes" with properties:
======================================
children: "Compound"

visible = "on"
axes_visible = ["on","on","on"]
axes_reverse = ["off","off","off"]
-->E=gce()         // handle Compound child of A
 E  =
Handle of type "Compound" with properties:
==========================================
children: matrix 41x1
visible = "on"
user_data = []
tag =

-->F.children(1).children(1)  // equal to E
 ans  =
Handle of type "Compound" with properties:
==========================================
children: matrix 41x1
visible = "on"
user_data = []
tag =
```

By following this method, you can retrieve or modify the properties of several handles with just one command, as shown in the following example:

```
-->clf;

-->plot();         // figure with two plots

-->F=gcf()         // handle Figure
 F  =
Handle of type "Figure" with properties:
========================================
figure_position = [200,200]
```

```
figure_size = [626,586]
axes_size = [610,460]
auto_resize = "on"
viewport = [0,0]
-->A=F.children(2)  // handle Axe
 A  =
Handle of type "Axes" with properties:
======================================
children: "Compound"

visible = "on"
axes_visible = ["on","on","on"]
axes_reverse = ["off","off","off"]
-->// A has three children

-->F.children(2).children
 ans  =
Handle of type "Compound" with properties:
==========================================
children: ["Polyline","Polyline","Polyline"]
visible = "on"
user_data = []
tag =

-->//  Three children of type Polyline

-->F.children(2).children.children.type
 ans  =
!Polyline  Polyline  Polyline  !

-->// get the three colors of the Polylines

-->F.children(2).children.children.foreground
 ans  =
    6.    1.    5.

-->// get only the last two colors

-->F.children(2).children.children(1:2).foreground
 ans  =
    6.    1.
```

Caution › *When a set of graphics command gets executed sequentially, this creates handles that are added to the graphics hierarchy in the same order and are stored inside a handle* ***H*** *of type* compound. *The property* `H.children` *is then a matrix of* `n` *handles where the last drawn entity corresponds to* `H.children(1)` *handle and the first to* `H.children(n)`.

You can find a complete description of the properties of each handle type in Scilab's online help. Enter `help graphics_entities` to get to it from the console. Each handle type has its own entry, for example, to see the description of `Figure` handles, you can call it directly from the console with the command `help figure_properties`. Table 19.1 gives a recap of the main graphics functions associated to the different handle types.

Table 19.1 : Recap of different types of handles

Handle	**Description**
Figure	Top graphics level, has one or several children that are Axes handles. Retrieve its handle with the command `get("current_figure")` or `gcf`.
Axes	Second graphics level, has children that are handles of different types but are often Compound handles. Retrieve its handle with the command `get("current_axes")` or `gca`.
Compound	Third graphics level. This entity is used to combine different graphics objects (with the help of the functions `glue` and `unglue`). Its children belong to types different from *figure*, *axes* and *compound*.
Handles to draw lines	
Polyline	Handle for lines drawn with `plot`, `plot2d`, `param3d`, `param3d1`.
Segs	Properties of segments drawn with `xsegs` and of arrows drawn with `xarrows`.
Arc	Handle of ellipses or ellipse arcs drawn with `xarc`, `xfarc`.
Rectangle	Handle of rectangles drawn with `xrect`, `xfrect` and `xrects`.
Champ	Display vector fields with the functions `champ`, `champ1` and `fchamp`.
Handles for plotting surfaces	
Plot3d	Handle of surfaces plotted with `plot3d`, `fplot3d`, `plot3d1` or `plot3d2`.
Grayplot	Handle of graphs of 2D surfaces displayed in gray or color with `grayplot` and `fgrayplot`.
Fec	Handle of graphs of 2D surfaces displayed in gray or color interpolated with `Sgrayplot` and `Sfgrayplot`.
Matplot	Integer matrices displayed with `Matplot` or `Matplot1`.
Figure text handles	
Text	Handle linked to text displayed with `xstring` or `xstringb`.
Legend	Handle of 2D curve legends (achieved with `legends` or `legend`).
Axis	Handle of coordinate axes drawn with `drawaxis`.
Label	Handle of labels associated to a plot's coordinate axes. Their parent is the Axes handle in which the corresponding plots were made. They do not appear inside the list of children of the Axes handle but are found inside its properties `x_label`, `y_label` and `z_label`!

Except for the Figure, Axes, Compound, Axis and Label handles which always occupy the same location in the graphics hierarchy, the other handles are either direct descendants of an Axes handle or are gathered inside a Compound handle.

*Tip › To retrieve the handle of a graphics entity, the best method consists in calling the command **E=gce()** right after the graphics command used to plot the entity. The entity's properties are directly accessible from the handle **E** or sometimes from its descendants.*

Each handle has the properties `tag` and `user_data` which let you add personalized user data to a graphics object.

```
-->surf();

-->F=gcf();                    // figure handle

-->F.user_data=[1:5];          // user data

-->F.tag="test figure";        // user tag

-->F  // handle content
 ans  =
Handle of type "Figure" with properties:
========================================
figure_position = [200,200]
figure_size = [626,586]
axes_size = [610,460]
auto_resize = "on"
viewport = [0,0]
```

This capability is very useful when one wants to create graphics interfaces (see Section 23.4, *Creating your own graphical interfaces*).

19.4. Working with several graphics windows

Scilab lets you work with several graphics windows at the same time. Each of these windows corresponds to a different Figure handle which you can label with a numerical index. This index is the number displayed in the window's upper-left corner and corresponds to one of the handle properties that defines the figure: `Figure_id` (see Figure 19.7). Two commands make it easier to manipulate windows:

- `winsid` lists the identifiers of existing graphics windows.
- `get_figure_handle` lets you then find a handle from its identifier.

Figure 19.7 : *Figure numerical identifier*

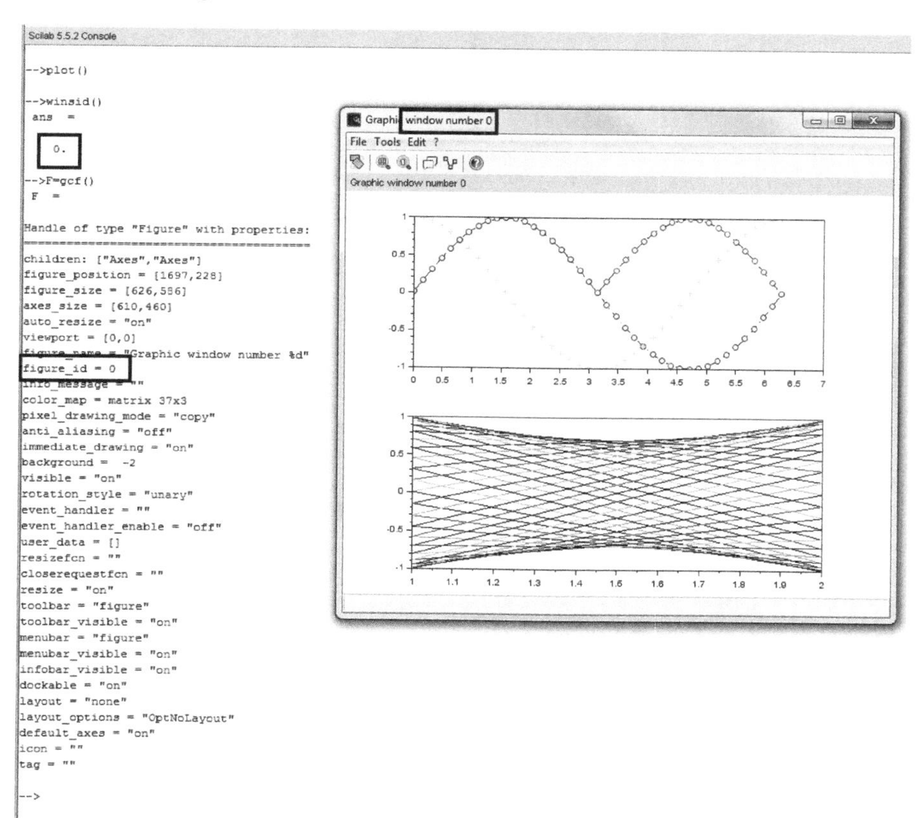

Commands used to manipulate different graphics windows take the handle of a figure as argument but also sometimes the numerical identifier of a figure:

- `clf` erases the figure's content given the figure's handle or its numerical index (and creates a blank figure if the identifier doesn't exist yet).
- `show_window` brings to the front the graphics window linked to its handle or numerical index.
- `delete` lets you delete a handle. If it is a Figure handle, the associated graphics window gets closed.

```
-->plot()       // figure 0

-->F0=gcf();  // figure 0 handle

-->A=gca();    // axes handle of figure 0

-->winsid()
 ans  =
    0.

-->F1=scf(1);    // generate an empty figure 1

-->delete(A)     // delete part of figure 0

-->winsid()
 ans  =
    0.    1.

-->surf()         // plot inside figure 1

-->clf(0)         // erase figure 0

-->winsid()
 ans  =
    0.    1.

-->delete(F0)    // delete figure 0

-->winsid()
 ans  =
    1.

-->delete(F1)    // delete figure 1

-->winsid()      // no more figures
 ans  =
     []

-->F1           // cannot access handle F1 anymore
 ans  =

    !--error 999

 get: The handle is not or no more valid.
```

Caution › *When several graphics commands are launched without erasing the current graphics window or without creating a new one, the new graphics entities overlap the ones that were already there while modifying the current axes properties.*

You can define or retrieve Figure or Axes handles by default by using the following commands:

- `sdf` or `set("default_figure")` to set the default figure *via* its identifier

- `gdf` or `get("default_figure")` to retrieve the current figure's handle
- `sda` or `set("default_axes")` to define the default axes *via* their handle
- `gda` or `get("default_axes")` to retrieve the current axes handle

After the commands `sdf` or `sda` are executed, all the new Figure or Axes entities use the format defined by `sdf` or `sda` as template.

To perform several plots inside the same figure, use the command `subplot`. This command lets you divide the figure in sub-figures arranged over `p` rows and `n` columns and places the plot number `k` in the corresponding sub-section with the command `subplot(p,n,k)` or `subplot(pnk)`. Each plot is in fact linked to an Axes handle who's parent is the Figure handle. For example, by executing the following script, you get the Figure 19.8.

```
clf()
subplot(221)
plot2d()
subplot(222)
surf()
subplot(2,2,3)
param3d()
subplot(2,2,4)
plot3d()
```

Figure 19.8 : *Several plots inside the same figure generated with* `subplot`

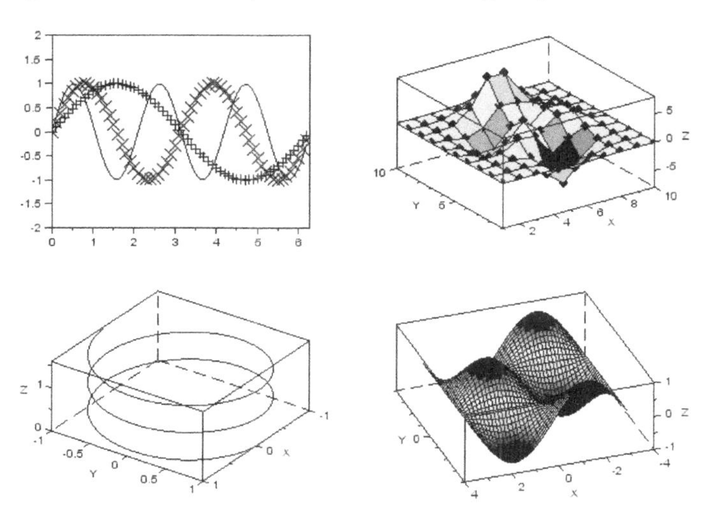

19.5. Exporting and saving plots

All figures created in Scilab can be exported as images of different formats by going to the graphics window's FILE menu and choosing:

- EXPORT TO for the formats PNG , JPG , GIF , BMP , PPM
- EXPORT TO VECTOR for the formats PS , EPS , PDF , SVG , EMF

You can also save images from the console or from a Scilab program by using the commands of the form `xs2*` where `*` is replaced with the chosen format:

- for matrix formats `xs2bmp`, `xs2gif`, `xs2jpg`, `xs2png`, `xs2ppm`
- for vector formats `xs2ps`, `xs2eps`, `xs2pdf`, `xs2svg`, `xs2emf`

These functions take as argument the window number of the figure to export along with a string containing the name of the file that the function needs to create. The image gets saved inside the current directory.

```
-->clf()                      // clear the current graphics window

-->surf()                     // new figure inside the current window

-->F=gcf();                   // figure handle

-->figure_id=F.figure_id  // graphics window number
 figure_id  =
    0.

-->xs2png(figure_id,'figure.png')  // export to png format

-->ls ('*.png')               // save inside the current directory
 ans  =
 figure.png
```

Caution › *The* **emf** *extension refers to the Microsoft Windows'* Enhanced Metafile *format. Exporting to this format is only available in Scilab's Windows version.*

Tip › *Since handles are variables, they can be saved inside a binary file, as with any other Scilab variable, by using the* **save** *function. You can later reload them into Scilab with the* **load** *command:*

```
clf;                       // clear the window
surf()                     // a figure
F=gcf();                   // handle of current figure
save('figure.sav','F')  // save F inside a file
delete(F)                  // delete the current window
load('figure.sav')         // load variable F
```

```
// the graphics window reopens and displays figure F
```

You can reload the saved plot with the **`load`** *command by using the* LOAD *tab under the* **`File`** *menu inside the graphics window.*

20

Two-dimensional Plot

Creating figures in the plane with Scilab is based on plotting segments. This simple geometrical structure is easily defined with a list of points, identified by their cartesian coordinates (often called x, y in what follows). Scilab's graphics functions take as arguments these lists of points stored inside vectors or matrices. They then display the corresponding segments as shown in Figure 20.1. The curves obtained correspond to Polyline handles; we will study some properties of these handles in this chapter.

Figure 20.1 : *Basics of plotting a curve in Scilab starting from coordinates of points* (x_i, y_i)

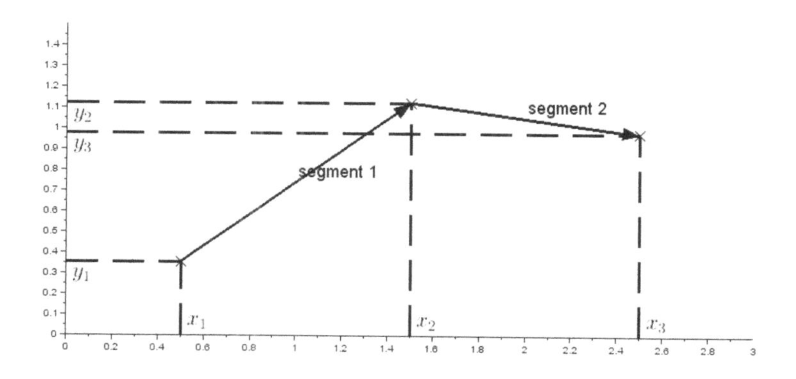

Caution › *Recall that in 2-D, objects drawn in Scilab are actually broken lines defined by vectors corresponding to coordinates of points and drawn by Polyline handles.*

20.1. Plotting with the *plot* command

To draw a curve in the plane, the most simple command you can use is the `plot` command. This function takes as input:

- two vectors x and y containing the list of horizontal and vertical coordinates of the points defining the broken line that needs to be drawn

- a string that describes the way in which the segments are to be plotted. It is made up of:
  - one letter defining the curve color (the first letter of the color's name in English *except for black* which is denoted by par `k`)
  - a combination of symbols indicating the marker placed on each point of the broken line and the line style

The color conventions, line styles and markers can be found in the following table.

Figure 20.2 : *Different types of markers, lines and colors for curves drawn with* `plot`

letter	k	b	g	c	r	m	y	w
color	black	blue	green	cyan	red	magenta	yellow	white

type	solid	dashed	dotted	dashed and dotted
symbol	-	–	:	-.

symbol	.	+	x	d	^	v	o	*	s	>	<	p
marker	•	+	×	◇	△	▽	○	*	□	▷	◁	☆

For example, execute the following script to draw the circle and line of Figure 20.3.

```
//*   file   testplot.sce   *//
clf; // clear the graphics window
// plot the circle  x(t)=cos(t), y(t)=sin(t)
t=[0:0.02:2*%pi];
x=cos(t);y=sin(t);
plot(x,y,'-r')
// plot the line y=sqrt(2)-x in blue
x=[0;sqrt(2)];y=[sqrt(2);0];
plot(x,y,'--b')
// plot the vector in cyan
x=1/sqrt(2)+[0;1];y=x;
plot(x,y,'c:o')
```

Figure 20.3 : Basics of plotting a curve in Scilab

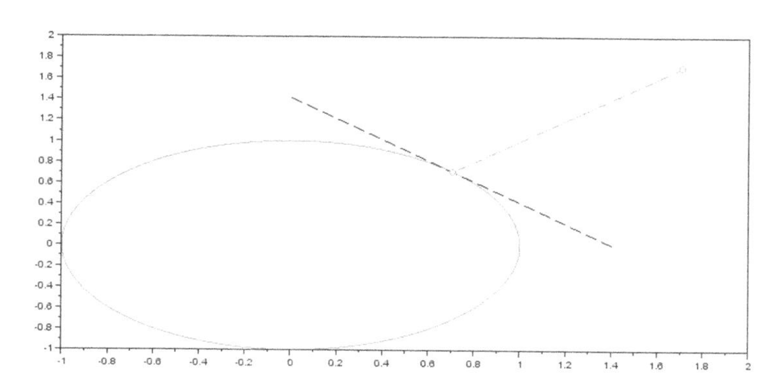

Tip › In the ***previous example*** *we constructed a vector* **t** *by using the increment operator ":", however we just as well could have used the* **linspace** *command to split the interval by providing the number of points rather than the increment step between each point.*

By modifying the value of the third argument in the `plot` function, you can easily get different types of lines for each curve. The different ways in which you can specify the type of lines to plot are better explained in the online help (enter `help LineSpec` in the Scilab console to get to it). You can also notice the connection between the curve aspects and the properties of the handles that define them. For example:

- The type of line drawn corresponds to the `line_style` property.
- The marker type corresponds to the `mark_style` property (see the list of codes Figure 20.14).
- The color used for the curve corresponds to the `foreground` property (see the list of codes Figure 20.10).

```
-->A=gca();          // figure axes

-->A.children(1).children.line_style  // 5 for dotted line
 ans  =
    5.

-->A.children(2).children.line_style  //  2 for dashes
 ans  =
    2.

-->A.children(3).children.line_style  //  1 for solid
```

```
 ans  =
    1.

-->A.children(3).children.foreground  //  5 for color red
 ans  =
    5.

-->A.children(1).children.mark_style  //  9 for o markers
 ans  =
    9.
```

Tip › Inside **Figure 20.3** *the circle looks oval! This is normal since we let Scilab set the scale as a function of the data sent by* `plot`*. The default scale is therefore not necessarily isometric. We can modify this behavior by changing the* `isoview` *property of the figure's Axes handle.*

```
-->A=gca();           // figure axes

-->A.isoview="on";  // change to isometric scale
```

After performing this modification you get the **following result***.*

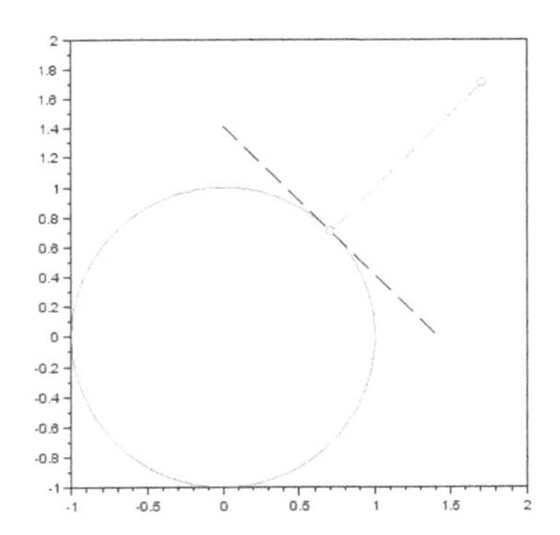

The `plot` function also has a different syntax from the one shown in the `testplot.sce` script. This makes it possible to avoid having to evaluate a function over a portion of the interval (here inside variable `x`):

```
function y=f(x)
    y=-sin(x^2)/x
endfunction
x=[0.001:0.02:2*%pi];
clf;
plot(x,f,"r")
```

Figure 20.4 : *Plotting a Scilab function with* `plot`

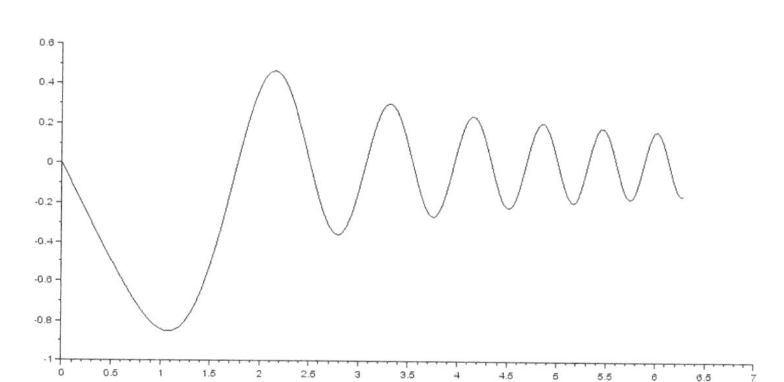

Caution › When a function is not accurately defined for a certain point, evaluating it for a vector can lead to an error. There are several possible solutions that you can implement to avoid this issue:

- *Shift the data to avoid having to evaluate the function at the point in question. This is shown in the* **previous example** *in the line* `x=[0.001:0.02:2*%pi]';` *which avoid the value* `0` *(we could also just have written* `x=%eps+[0:0.02:2*%pi]';`*), but this involves knowing which values are problematic in advance.*
- *Change Scilab's default behavior by using the command* **`ieee`** *so that an error is not created, for example when dividing by zero as shown below (also see* **example** *in* **Section 7.2, Elementary mathematical functions**).

```
function y=f(x)
    y=-sin(x^2)/x
endfunction
ieee(2)  // avoid division by 0
x=[0:0.02:2*%pi];
clf;
plot(x,f,"r")
```

Finally, to plot terms of a sequence (u_k) contained inside a vector, you may omit the horizontal coordinate of points to plot (they will default to the vector's index value for the data).

```
clf;
// computing terms of a sequence defined by
// u(k+1)=sin(u(k))   and u(1)=1
u(1)=1;
for k=1:20
    u(k+1)=sin(u(k));
end
```

```
// plot terms of the sequence
plot(u,'-*r')
```

Figure 20.5 : Plotting terms of a sequence contained inside a vector

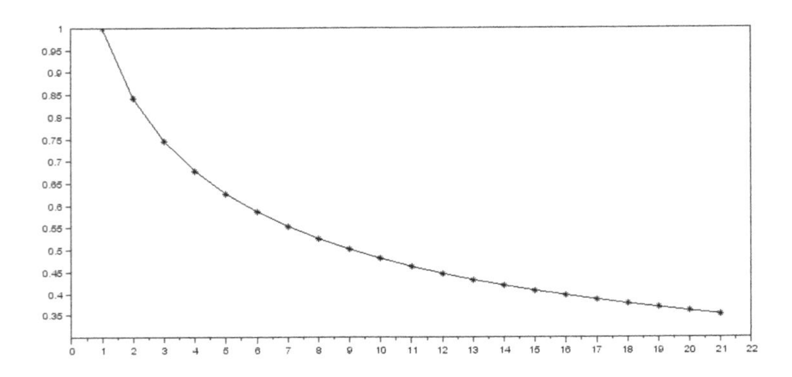

Zooming

You can zoom in on a part of a figure by using the `zoom_rect` command. You define the zone you wish to zoom on as a matrix of coordinates `[xmin,ymin,xmax,ymax]`. You can then zoom out with the command `unzoom`. These commands act on the current axes (the handle returned by `gca()`) and, in effect, consists in modifying the values of the `zoom_box` property. Execute the following script to get Figure 20.6.

```
-->exec('testplot.sce',-1)

-->A=gca();A.zoom_box
 ans  =
     []

-->zoom_rect([0.5,0.5,1,1]);              // zoom

-->A.zoom_box
 ans  =
    0.5    0.5    1.    1.  - 1.    1.

-->unzoom()                               // revert to initial plot

-->A.zoom_box
 ans  =
     []

-->A.zoom_box=[0.5,0.5,1,1];     // equivalent to zoom_rect
```

```
-->A.zoom_box
 ans  =
    0.5    0.5    1.    1.  - 1.    1.

-->unzoom()                               // revert to initial plot

-->A.zoom_box
 ans  =
     []
```

Figure 20.6 : Zooming with the `zoom_rect` function

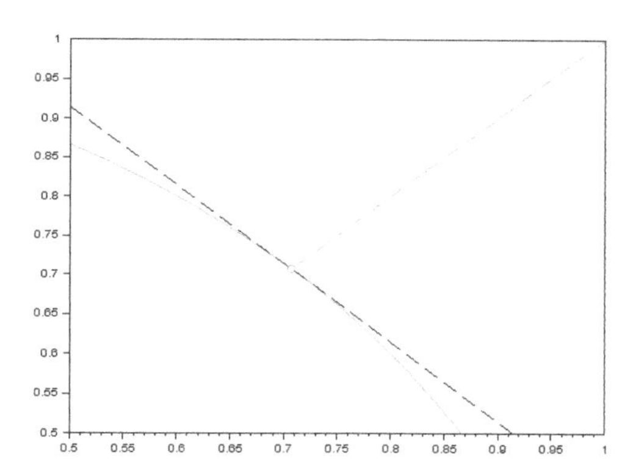

You can get a similar result to the one in Figure 20.6 by using the `replot` function with the same arguments as `zoom_rect`. To see this, you can look at the effect this command has on the Axes handle's properties.

```
-->exec('testplot.sce',-1)

-->A=gca();data_bounds=A.data_bounds,
 data_bounds  =
  - 0.9999987  - 0.9999710
    1.7071068    1.7071068

-->replot([0.5,0.5,1,1]);                 // modify A. data_bounds

-->A.data_bounds
 ans  =
    0.5    0.5
    1.     1.

-->replot(data_bounds');                  // revert to initial plot

-->A.data_bounds
 ans  =
  - 0.9999987  - 0.9999710
```

```
        1.7071068    1.7071068
```

The `replot` function therefore modifies the `data_bounds` property of the Axes handle.

Tip › *You can also zoom in on a figure by using:*

- *the mouse scroll (scroll forward and backward to zoom in and out, respectively)*
- *the graphics window's zooming tools:*
  - *from the TOOLS menu in the menu bar*
  - *the two Toolbar icons ZOOM AREA/ORIGINAL VIEW (see figure **below**)*

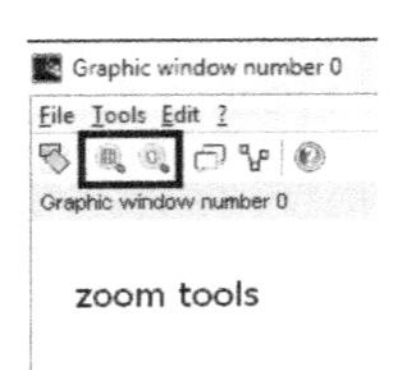

Caution › *The functions **`zoom_rect`** and **`replot`** take as input a vector with four columns `[xmin,ymin,xmax,ymax]` which defines the bounds of the graphics window. However, this data format is not compatible with the one used by the **`data_bounds`** property! To modify **`data_bounds`**, you need to input arguments as a two-by-two matrix `[xmin,ymin; xmax,ymax]`. On the other hand, the format `[xmin,ymin,xmax,ymax]` can be used to modify the **`zoom_box`** property. Keep in mind that Scilab sometimes adds the data **`zmin,zmax`** (corresponding to the third coordinate) to make this field compatible with 3D figures (see Chapter **Three-dimensional Plots**). You will also notice that it is easy to go from one format to another by using Scilab's matrix operations as shown below.*

```
-->rect=['xmin' 'ymin';'xmax' 'ymax']  //  format data_bounds
 rect  =
!xmin  ymin  !
!            !
!xmax  ymax  !

-->rect(:)'                            //  format zoom_box
 ans  =
!xmin  xmax  ymin  ymax  !

-->box=['xmin' 'ymin' 'xmax' 'ymax']  //  format zoom_box
 box  =
!xmin  ymin  xmax  ymax  !

-->[box(1:2);box(3:4)]                //  format data_bounds
 ans  =
!xmin  ymin  !
!            !
!xmax  ymax  !
```

Tip › *Since Scilab only plots broken lines, you need to play with the curve's point spacing to get a smooth curve rendering. If you zoom in excessively, you will start to notice the broken lines that make up the* *curve. However, the smaller the point spacing, the greater the number of points necessary to create the plot. This can increase the computation and plotting time. To notice this, execute the following commands for different values of* **dx** *(see the figure* **below***).*

```
clf;
dx=0.5;             // point spacing
x=[0:dx:%pi];       // horizontal coordinates
y=sin(x);           // vertical coordinates
plot(x,y,'-r')      // curve plot
```

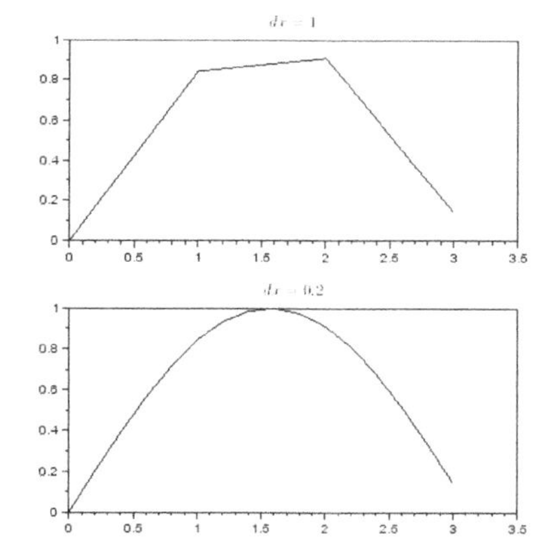

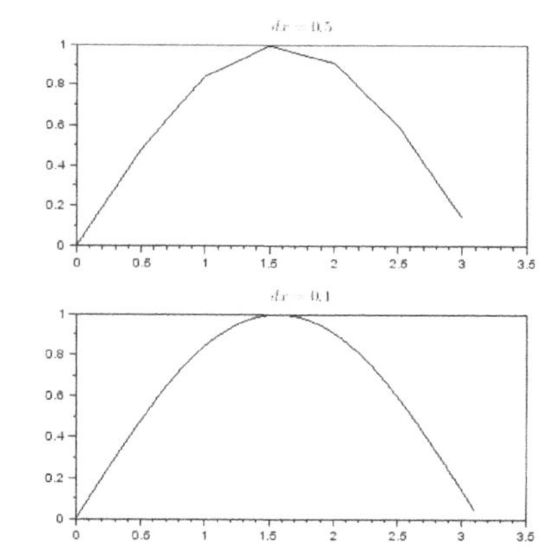

Caution › *Vector operations are used to determine* y *coordinates given* x *coordinates. The basic arithmetic operations* +, -, *, /, ^ *can then lead to some confusion regarding element-wise operations and matrix operations (see* **Section 9.2, Element-wise and matrix operations***). To avoid having to specify which ones are element-wise operations in the expressions used (by adding* . *to each operator), you will benefit from the use of the function* **feval** *which lets a function act on each element of a matrix or vector (see* **Section 16.1, Defining a function***).*

```
-->function y=f(x)
-->    y=sin(x)/x
-->endfunction


-->x=[-2*%pi:0.02:2*%pi]';size(x)  // x coordinate
 ans  =
    629.    1.

-->y=f(x);size(y)                   // y is not the correct size
 ans  =
    629.    629.

-->y=sin(x)./x; size(y)             // y is the correct size
 ans  =
    629.    1.
```

```
-->y=feval(x,f); size(y)          // y is the correct size
 ans  =
    629.    1.
```

Plotting several curves with one command

You can plot several curves with one plot command by concatenating the instructions for each curve as shown in the following example:

```
// line
X=[-2;4];Y=2*X-1;
// point scatter
x=4*grand(100,1,'def')-1; // x coordinates
noise=(2-abs(x-1)).*(2*rand(x)-1);
y=2*x-1+noise // y coordinates
// plot
clf;
plot(X,Y,'-r',x,y,'+k')
```

The first plot is performed with one solid red line (due to the '-r' argument) and the second with black + markers (since it uses the '+k' argument), which yields the result in Figure 20.7.

Figure 20.7 : *Several plots with one graphics command*

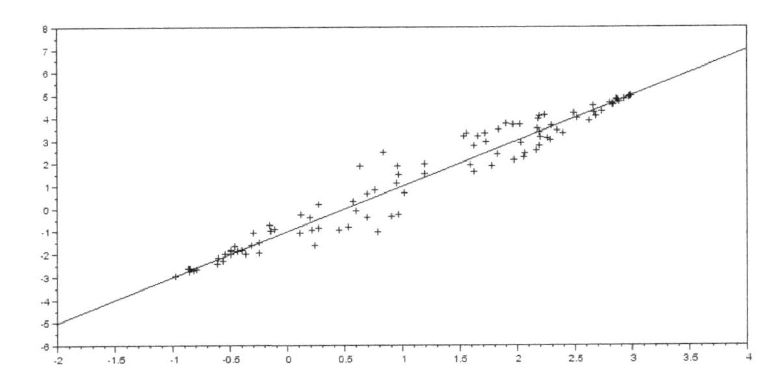

20.2. Titles, grids, legends and colors

To improve the readability of your plots, you can easily add various elements such as:

- a title with the command `title` or `xtitle` (also see section *Adding a title or title page*)
- a coordinate grid with `xgrid`
- legends with `legends`, `legend` or `captions`

Execute the following script to get Figure 20.8.

```
exec('testplot.sce',-1)
// add a title
xtitle(' figure title')
// add a green (=3) coordinate grid
xgrid(3)
// add a legend in the upper-right corner (2) with a box (1) surrounding
 it
legend('circle','tangent','normal',2,1);
// add a legend in the lower-right corner (4) without a box (0)
legends(['circle','tangent','normal'],[5,2,4],4,0);
```

Figure 20.8 : *Adding a title, coordinate grid and legend*

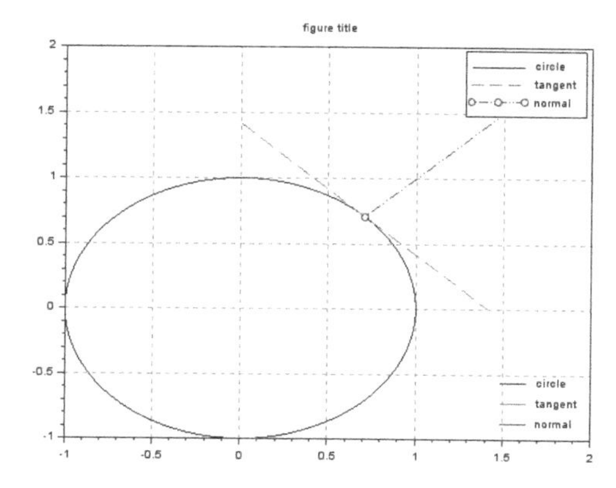

Adding labels

If you'd prefer to add a label to the different curves instead of a legend, use *datatips*. You can create labels with the command `datatipCreate` by specifying the curve's handle and the curve point at which it needs to be labeled. There are several ways to define the point, such as with:

- a matrix `[x,y]` containing the point's coordinates
- an index `i` corresponding to the point's coordinates `x(i),y(i)` belonging to the curve created with the command `plot(x,y)`

The label displayed is a Compound handle made up of two components: one for the text and another for the label contour line (of type *polyline*). Execute the following script to get Figure 20.9.

```
clf;
ieee(2)  // to warn of division by 0
x=[0:0.02:2*%pi];
y1=cos(x);y2=sin(x);y3=-sin(x.^2)./x;
plot(x,y1,'-b',x,y2,'-g',x,y3,'-r')
E=gce();
// datatip for the first curve
P1=E.children(1);
T=datatipCreate(P1,100);
// coordinates of 100th point
x(100),y3(100)
// datatip for the second curve
P2=E.children(2);
T=datatipCreate(P2,[x(200),y2(200)]);
T.type      // datatip handle
T.children  // Polyline and Text
```

Figure 20.9 : *Adding labels to curves*

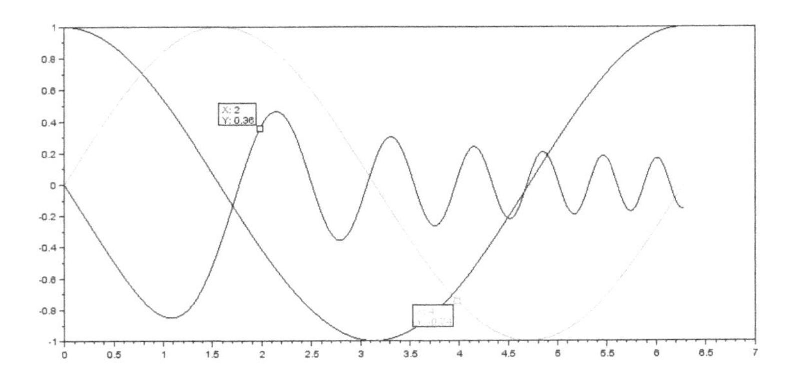

You can check from the console if the coordinates of points attached to labels match the rounded values displayed in the figure:

```
-->clf;

-->ieee(2)  // to warn of division by 0

-->x=[0:0.02:2*%pi];

-->y1=cos(x);y2=sin(x);y3=-sin(x.^2)./x;

-->plot(x,y1,'-b',x,y2,'-g',x,y3,'-r')

-->E=gce();

-->// datatip for the first curve

-->P1=E.children(1);

-->T=datatipCreate(P1,100);

-->//coordinates of 100th point

-->x(100),y3(100)
 ans  =
    1.98
 ans  =
    0.3547632

-->// datatip for the second curve

-->P2=E.children(2);

-->T=datatipCreate(P2,[x(200),y2(200)]);

-->T.type       // datatip handle
 ans  =
 Datatip

-->T.children  // Polyline and Text
 ans  =
     []
```

If this way of creating labels does not suit you, you can also use the graphics window's EDIT menu and go to the sub-menus START DATATIP MANAGER and STOP DATATIP MANAGER to edit labels. The graphics window then calls the functions `datatipToggle()` and `datatipMove()` which let you set and move *datatips* by using just the mouse.

Color management

The `xgrid` and `legends` functions take as input a number which specifies colors. These numbers are then used by the `foreground` and `background` properties to define the colors of a Polyline handle. They correspond to the color table in Figure 20.10 (see the warning note that follows).

Figure 20.10 : Default color table

number	1	2	3	4	5	6	7	8
color	black	blue	green	cyan	red	magenta	yellow	white

You can manage curve colors in different ways with Scilab:

- *via* a number (by default those given in Figure 20.10)
- *via* a string (the color name)
- *via* the RGB color coding format

The functions `color`, `name2rgb` and `rgb2name` let you create links between these three descriptions:

```
-->//red = rgb color (255,0,0)

-->rgb=name2rgb("red")
 rgb  =
    255.    0.    0.

-->rgb2name(rgb)
 ans  =
!red   !
!      !
!red1  !

-->// red = color number 5

-->color("red")
 ans  =
    5.

-->color(rgb(1),rgb(2),rgb(3))
 ans  =
    5.
```

Caution › You also have the possibility of changing the color table used by Scilab (see **Section 21.3, Facets and surfaces Figure 21.10***). In this case, the color numbers listed in the table in* **Figure 20.10** *are no longer valid! It's better to use the* **`color`** *command to specify a color by its name, even if it is longer to write.*

You can find the list of colors in the online help `help color_list`. If you prefer choosing your colors *via* a graphics interface, use the commands:

- `getcolor` to get the Scilab number and color name (see Figure 20.11)
- `uigetcolor` to get a color's RGB (see Figure 20.12)

Figure 20.11 : Graphics interface to select a color `getcolor`

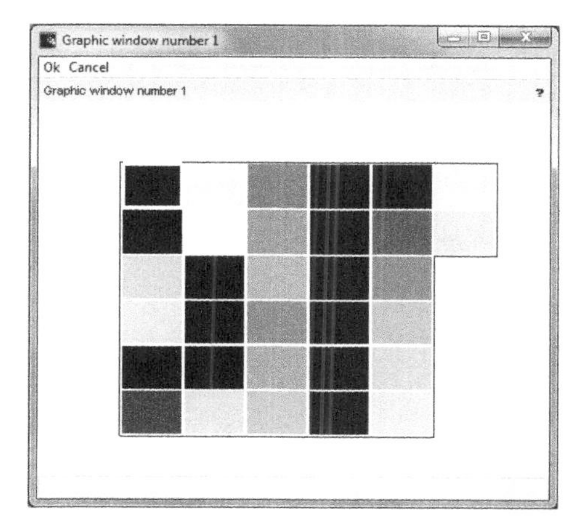

Figure 20.12 : Graphics interface to select a color `uigetcolor`

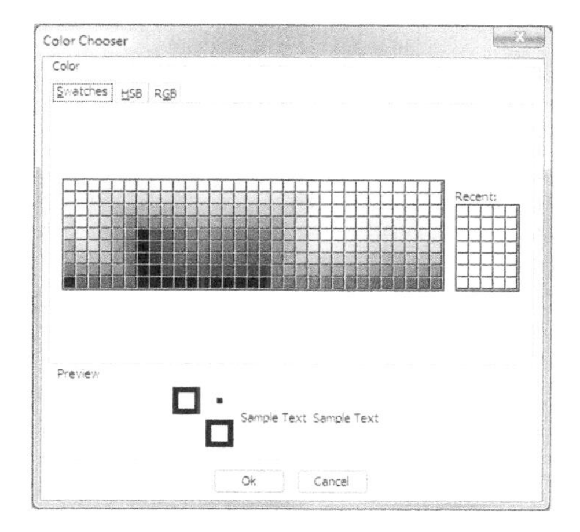

Changing the environment color

The function `colordef` lets you change several parameters of the graphics environment. For example, you can plot on a black background (with the argument `'black'`) instead of a white background. If you execute the script used to demonstrate the `plot` command again after using the command `colordef('black')`, you will get the same curves as in Figure 20.3, but with a black background (see Figure 20.13).

```
colordef('black')          // plot curves on a black background
exec('testplot.sce',-1)
colordef('white')          // to revert to the default mode
// but you still need to redefine the background color as white
id_white=addcolor(name2rgb("white")/255);  // add white
F=gcf();
F.background=id_white;     // assign it to the background color
sdf()                      // new default figure
```

Figure 20.13 : *Plotting on a black background with* `colordef`

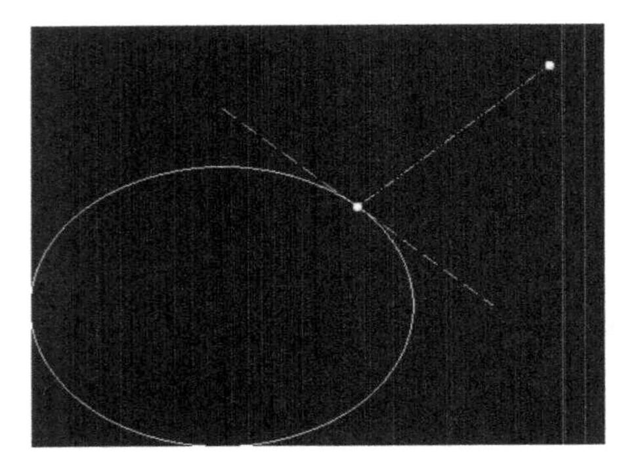

*Caution › **`colordef`** modifies the default figure and axes styles for all the subsequent graphics window since the command calls **`sdf`** and **`sda`**. To revert to the default style, use the command **`colordef('white')`**. If the background color does not return to white, this means white is no longer present in the figure's **`colormap`**. In this case, you need to add this color with the command **`addcolor`**, then assign it to the background color and save the default figure template with **`sdf()`**.*

Marker management

Managing markers can also be performed with the use of a number instead of the symbols `o,<,>,*,...` used by the `plot` command (see Figure 20.14). These numbers are the ones used by the `mark_style` property in the Polyline handles.

Figure 20.14 : *List of available markers for plots*

k	0	1	2	3	4	5	6	7	8	9	10	11	12	13	14
marker	•	+	×	⊕	♦	◇	△	▽	◈	○	∗	□	▷	◁	☆

Plotting error bars

In order to plot error bars around points in a curve, use the command `errbar`. This function takes as input the matrices `x,y` which define the curve as well as two matrices em, ep which define, for each point in the curve, the error margin as a distance above and below the point. You can use the following script to get the plot in Figure 20.15.

```
clf;
ieee(2) // to warn of division by 0
// divide the interval into 125 points
x=linspace(0,2*%pi,125);
y1=x.*cos(x);
plot(x,y1,'-b')
// error margin
em=grand(x,'def');    // distance below
ep=grand(x,'def');    // distance above
errbar(x,y1,em,ep)     // plotting bars
// error bar handle
E=gce();
E.type          // Segs
E.segs_color=color("blue")  // color blue
// second curve
y2=%pi*sin(x)./x;
plot(x,y2,'-r')
// error margin
em=ones(x);ep=em;
errbar(x,y2,em,ep)     // plotting bars
E=gce();E.segs_color=color("red")  // color red
```

The error margins in the first curve were randomly computed whereas, for the second curve, they were set at plus or minus one. The errbar function creates entities of type segs which are covered in more details later on in Section 22.4, *Arrows and segments*.

Figure 20.15 : *Plotting Error bars with errbar*

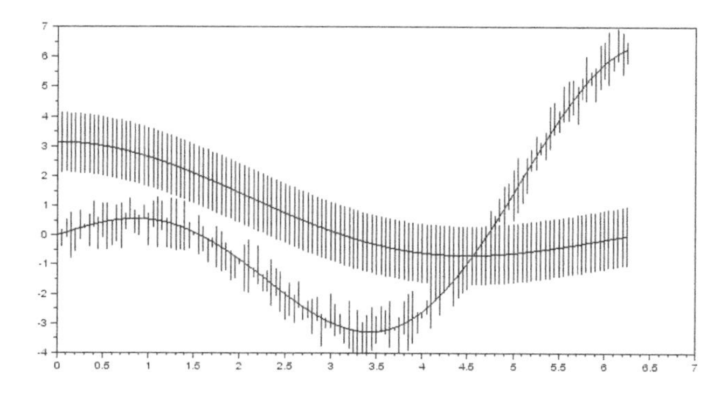

20.3. *plot2d* command and other types of plots

For more advanced features, you will need to use the `plot2d` command: it provides more plotting options than `plot` at the expense of a more complex syntax. The basic format is `plot2d(x,y,c)` where `x` and `y` are vectors that define the x and y coordinates of the curve points and `c` is a number that sets the color used for the plot (see Figure 20.10). If you wish to plot several curves with the same exact number of points, you can plot them with just one command by concatenating (with `[ ]`) the different vectors that define the curves to plot. For example, the script:

```
clf;
x=[0.001:0.02:2*%pi]';
y1=cos(x);y2=sin(x);y3=-sin(x.^2)./x;
plot2d([x x x],[y1 y2 y3],[2 4 5])
```

yields the three curves in Figure 20.16.

Figure 20.16 : *Plotting several curves with* `plot2d`

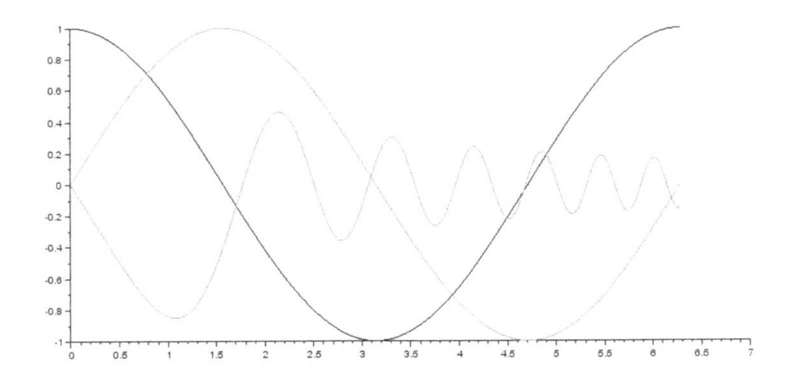

Caution › To make sure **`plot2d`** *indeed plots three distinct curves as shown in the example in* **Figure 20.16**, *the vectors containing the coordinates of the curve's points need to be* column vectors. *If on the other hand you have the opposite situation,* `[x,x,x]` *will be a matrix of only one row instead of a matrix of three columns and* **n** *rows, and therefore seen as only one curve!*

Tip › You can ask Scilab to perform a plot with markers rather than a solid line by entering a negative value inside the third argument **c**. *In this case, the marker corresponding to the number* **-c** *will be used for the plot (see* **Figure 20.14**).

As a result, the script:

```
clf;
// color plot
x=[0.001:0.02:2*%pi]';
y1=cos(x);y2=sin(x);y3=-sin(x.^2)./x;
plot2d([x x x],[y1 y2 y3],[2 4 5])
// plot with markers
x=[0.001:0.2:2*%pi]';
y1=cos(x);y2=sin(x);y3=-sin(x.^2)./x
plot2d([x x x],[y1 y2 y3],[-1 -3 -5])
```

will yield the following three curves:

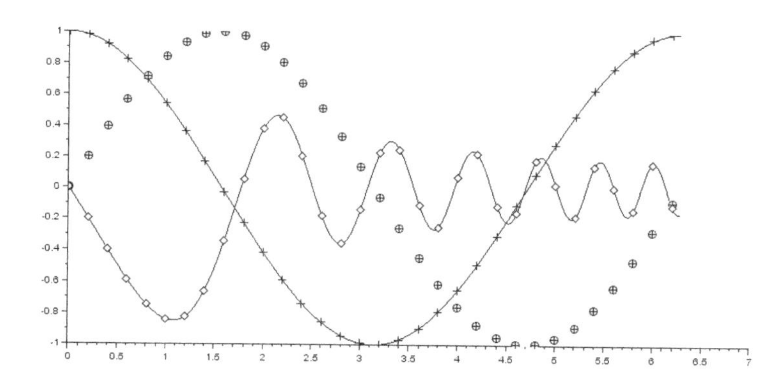

The optional parameters of `plot2d` are specified in the form of `key=value`. The most useful ones are:

- `rect` lets you define the extreme limits of the x and y coordinates to display inside the graphics window. The format used is `rect=[xmin, ymin, xmax, ymax]`.
- `axesflag` manages the axes location, for example:
  - `axesflag=0` does not add axes or a box around the plot
  - `axesflag=2` to display only a box around the plot
  - `axesflag=1` or `3` to change the `y` axis position
  - `axesflag=4` or `5` to intersect the axes in the center of the figure
- `frameflag` manages the window size and the scale:
  - `frameflag=0` to superimpose a graph onto the previous window while keeping the scale
  - `frameflag=1` or `2` for a scale computed from data provided with the option `rect`

- `frameflag=2` or `4` for a scale computed from the minima/maxima of the curves' points
- `frameflag=3` or `4` for an isometric scale
- `frameflag=5` or `6` for a scale with simplified ticks

Tip › There is no option to plot a curve with coordinate axes that intersect at the `(0,0)` point in the `plot` and `plot2d` commands. To achieve this, you will need to modify the plot's `x_location` and `y_location` properties as follows:

```
clf;
t=[0:0.02:10*%pi]';  // matrix of parameters
// warning, needs the element-wise operator .* :
x=sin(t).*t;y=cos(t).*t;
plot(x,y,'-r')          // plot
// intersect the coordinate axes at point (0,0)
A=gca();
A.x_location="origin";
A.y_location="origin";
```

which yields:

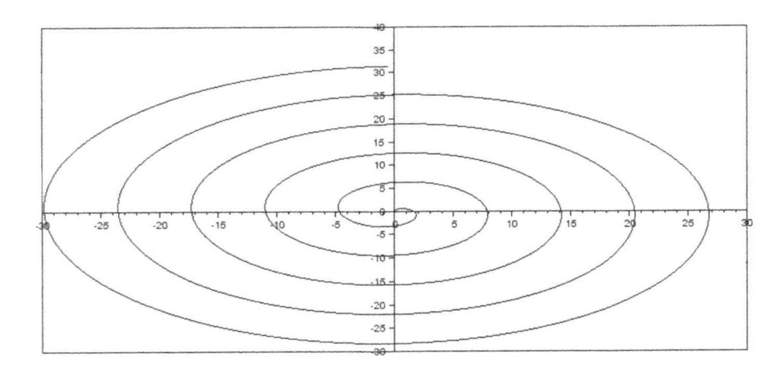

Curve of type $y = f(x)$

If you need to plot a curve of the form $y = f(x)$, use `fplot2d`. This command lets you avoid computing the y coordinates with `feval`. This is similar to what `plot` lets you do with the syntax seen previously. For example, execute the script:

```
function y=f(x)
    y=-sin(x^2)/x
endfunction
ieee(2) // to warn of division by 0
```

```
x=[0:0.02:2*%pi];
clf;
fplot2d(x,f,5)
```

to get the result Figure 20.17.

Figure 20.17 : *Plotting a Scilab function with* `fplot2d`

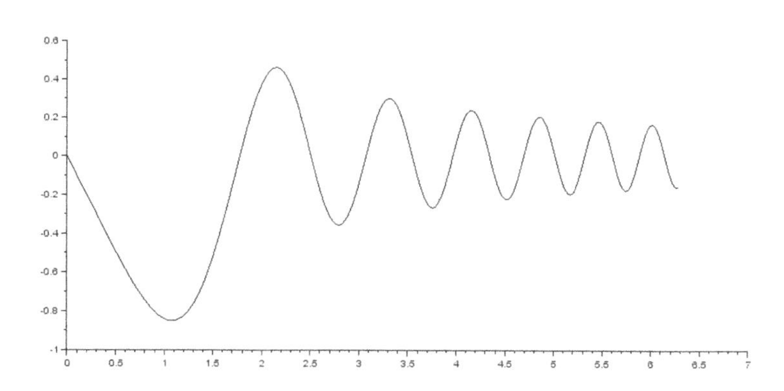

Plotting curves as step functions, vertical bars or arrows

The `plot2d` command has several versions which can be called by following the command format `plot2d*`. It lets you generate different types of curves, each corresponding to a different value of the `polyline_style` property:

- `plot2d2`: a step function plot corresponding to `polyline_style=2`
- `plot2d3`: a vertical bar plot corresponding to `polyline_style=3`
- `plot2d4`: a plot of arrows corresponding to `polyline_style=4`

Execute the following script to get the result in Figure 20.18.

```
x=[0:0.4:%pi]';y=sin(x);
clf()
subplot(221)
plot2d2(x,y,5)
xtitle('plot2d2 <=> polyline_style=2')
subplot(222)
plot2d3(x,y,5)
xtitle('plot2d3 <=> polyline_style=3')
subplot(223)
plot2d4(x,y,5)
```

```
xtitle('plot2d4 <=> polyline_style=4')
// increase the arrow size in plot2d4
E=gce();
E.children.children.arrow_size_factor=3;
subplot(224)
xtitle('plot2d with polyline_style=5')
plot2d(x,y,5)
// increase the arrow size in plot2d4
E=gce();
E.children.polyline_style=5;
```

You can get the same types of plots by modifying the `polyline_style` property.

Figure 20.18 : *Other versions of* `plot2d`

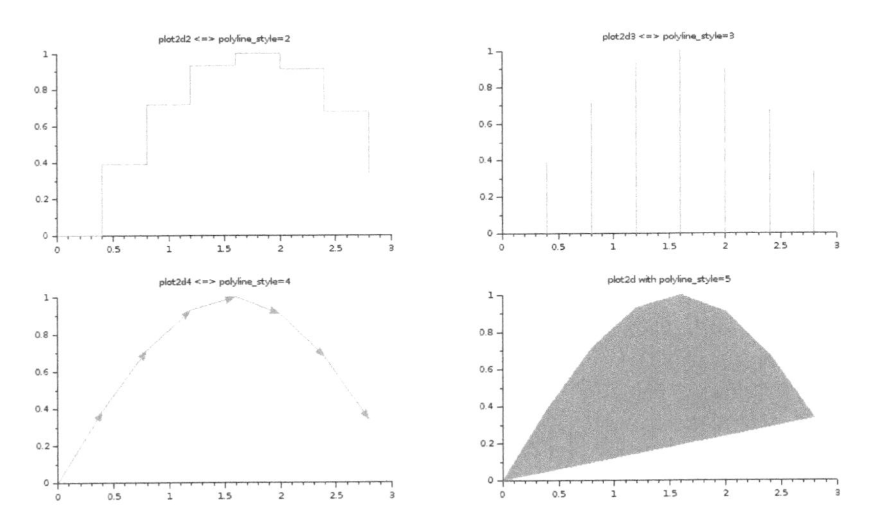

Curves defined by a polar equation

You can plot curves defined by equations in polar coordinates by displaying the suitable coordinate system with the `polarplot` command. Execute the following script:

```
clf()
theta=[0:0.02:2*%pi]';
r=1+0.2*cos(theta.^2);
polarplot(theta,r,style=5);
// axes modification
a=gca();a.isoview='on'
// decrease the window size
```

```
a.data_bounds=[-1.2,-1.2;1.2,01.2]
```

You will get Figure 20.19 with a plot of the curve defined by the equation $r(\theta)=1+0.2\times\cos(\theta^2)$ where θ ranges from 0 to 2π radian. Axes drawn by `polarplot` let you visualize the distance r to the origin and the angle θ *which is displayed in degrees and not radians!*

Figure 20.19 : *Curve in polar coordinates achieved with* `polarplot`

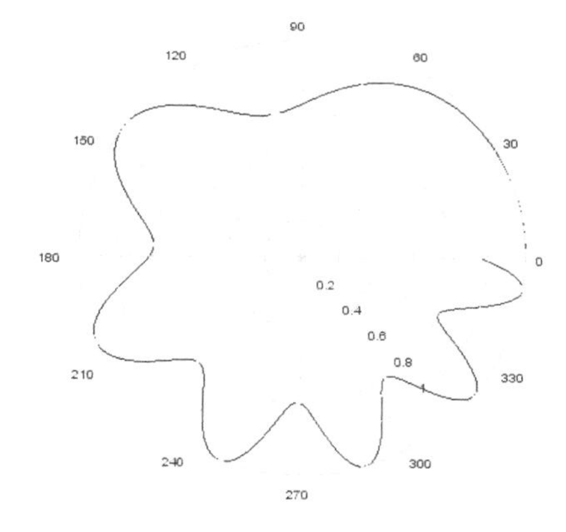

21 Three-dimensional Plots

Creating figures in 3D space with Scilab is based on the definition of points that are analogous to those used in a plane, but that have three coordinates (usually denoted x, y, z) instead of two.

21.1. View angle

The main difference compared to a planar plot is that in order to display a 3D plot on the screen, the figure needs to be projected onto a plane (the plane of your computer's screen). The result will strongly depend on the angle at which the figure is viewed. This view angle is defined by two values (from now on denoted α and θ) which are angles expressed in degrees. You will often need the help of the diagram in Figure 21.1 to pick the correct angle values that define the view origin you wish to use.

Figure 21.1 : Viewing point as a function of angles α and θ

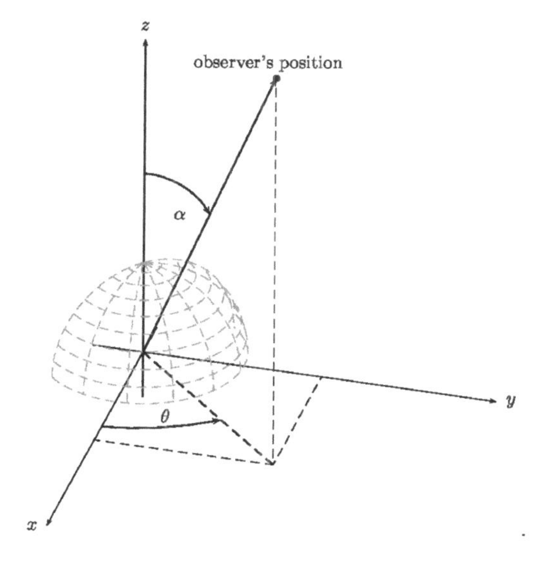

This data is stored inside the Axes handle's `rotation_angles` property for a given plot. For example, for a figure displayed by using the `surf()` command (see Example 19.1), these angles are `alpha=51` and `theta=-125`, respectively.

```
-->clf;

-->surf()

-->A=gca();

-->A.rotation_angles
 ans  =
    51.  - 125.

-->alpha=A.rotation_angles(1)
 alpha  =
    51.

-->theta=A.rotation_angles(2)
 theta  =
  - 125.
```

Most of Scilab's 3D plotting functions accept optional inputs of the format `alpha=value` and `theta=value` to define the view point for the way the figure gets displayed. To understand the link between these values and the view direction, modify the values of the `rotation_angles` property and check that:

- `theta` can range from `0` to `360` degrees with:
  - `theta=0` view from the direction `x=0` and `y>0`
  - `theta=90` view from the direction `x>0` and `y=0`
  - `theta=180` view from the direction `x=0` and `y<0`
  - `theta=270` view from the direction `x<0` and `y=0`

Figure 21.2 : Appearance of figure `surf()` for different values of `theta`

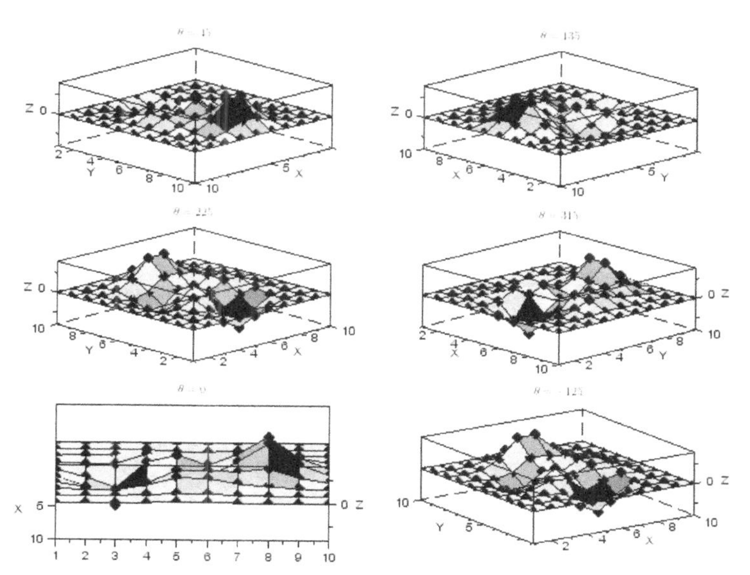

- `alpha` can range from `0` to `180` degrees with:
  - `alpha=90` view from the `z=0` plane
  - `alpha<90` view from a point of coordinate `z>0` (meaning "above" the `z=0` plane)
  - `alpha>90` view from a point of coordinate `z<0` (meaning "below" the `z=0` plane)

Figure 21.3 : Appearance of figure `surf()` for different values of `alpha`

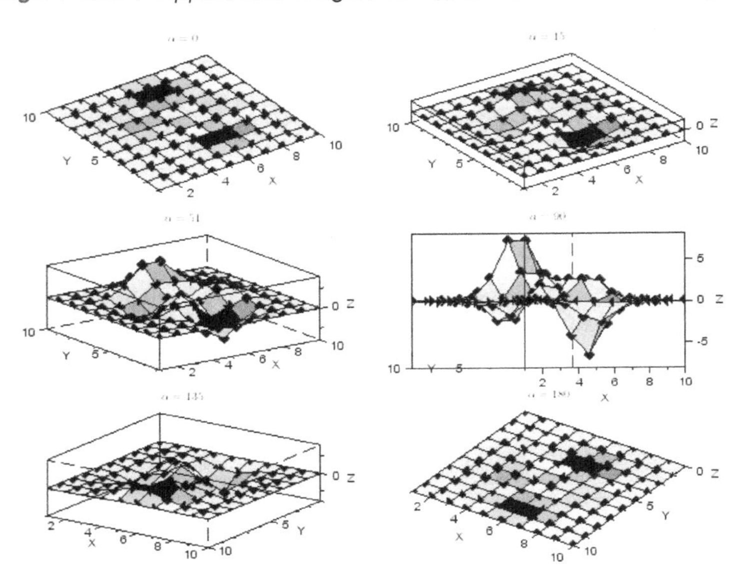

Tip › You can modify the viewing direction by using:

- *the mouse (keep the right mouse button pressed, then move the mouse to rotate the figure)*
- *the Rotate tool in the graphics window (see the figure* **below***):*
  - *from the TOOLS menu in the menu bar*
  - *the Rotate icon in the toolbar*

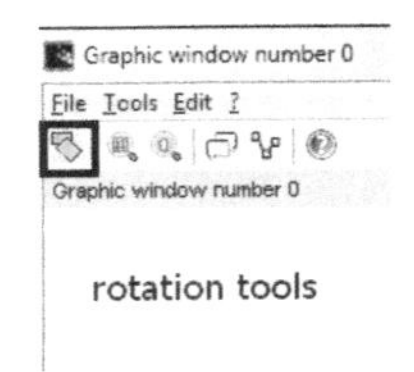

Zooming is performed the same way as for planar figures.

Caution *› The values of* **`theta`** *and* **`alpha`** *are angles (in degrees) and as such are defined with the modulo* ***360****.*

21.2. Curves in 3D space

To plot curves in 3D space, you will use the command `param3d`. For example, execute the following script to get the helix displayed in Figure 21.4.

```
function [x,y,z]=helix(t)
    x=cos(t)
    y=sin(t)
    z=t
endfunction
// compute coordinates of points
t=[-5*%pi:0.02:5*%pi];
[x,y,z]=helix(t);
// display the curve
clf;
param3d(x,y,z,alpha=15,theta=50)
E=gce();E.foreground=5 // modify the curve's color
```

The coordinate of points of the helix are calculated with the `helix` function, from a vector of values of the parameter `t` and are then displayed with `param3d`. To modify the curve's appearance, you need to retrieve the current handle (of type *polyline*) right after calling `param3d`. Here, we have assigned the number corresponding to the color red to the `foreground` property.

Figure 21.4 : *Plotting a curve in 3D space with* `param3d`

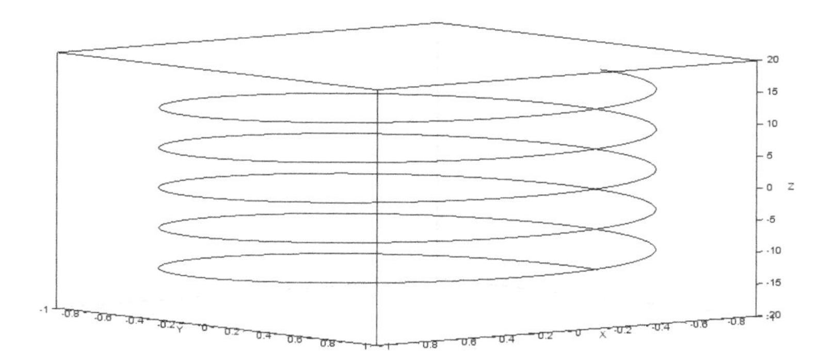

If you wish to specify the style of the curve to use, it is preferable to use the command `param3d1`. The syntax is a little more complicated: you need to replace the `z` coordinate with `list(z,style)` where style is the number which indicates the color or marker to use for the plot. For example, execute the following script to get Figure 21.5.

```
function [x,y,z]=helix(t)
    x=cos(t)
    y=sin(t)
    z=t
endfunction
// compute coordinates of points
t=[-5*%pi:0.02:5*%pi];
[x,y,z]=helix(t);
// display the curve
clf;
param3d(x,y,z,alpha=15,theta=50)
E=gce();E.foreground=5 // modify the curve's color
// display points marked by a "O"
t=10*%pi*grand(30,1,'def')-5*%pi;
[x,y,z]=helix(t);
param3d1(x,y,list(z,-9),alpha=15,theta=50)
```

Note that the `-9` in `list(z,-9)` corresponds to markers `o` according to the table in Figure 20.14.

Figure 21.5 : *Plotting a curve in 3D space with* `param3d1`

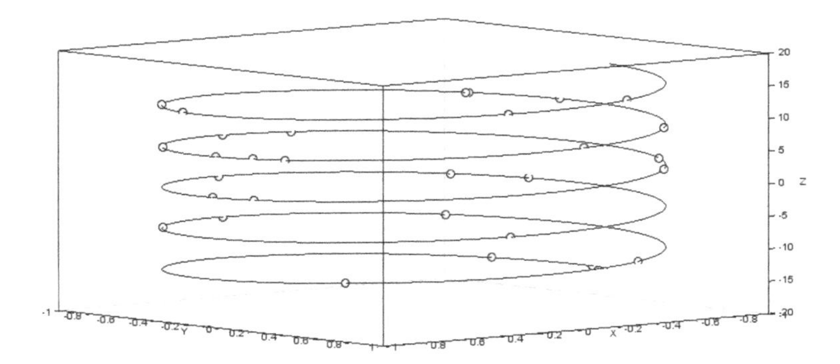

21.3. Facets and surfaces

Plotting surfaces in 3D with Scilab is based on drawing facets, which play the same role as segments for planar figures. This geometric structure is defined by a list of points and their associated cartesian coordinates. Scilab's graphics functions take as argument these lists of points stored inside vectors or matrices which they use to display the corresponding facets as in Figure 21.6.

Figure 21.6 : Plotting a facet with Scilab from points of coordinates (x_i, y_i, z_i)

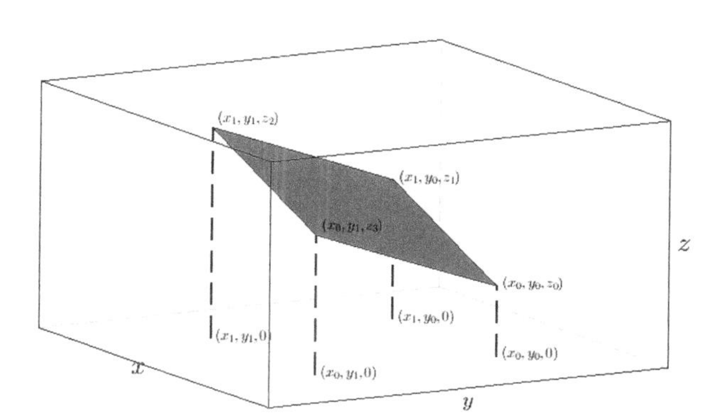

Caution › *A facet has two faces: one above and another below. The face oriented towards the positive z direction is the upper face if, when looking at the facet from that direction, rotating in the positive direction (counter-clockwise), the points delimiting the facet in* **Figure 21.6** *are ordered as follows:*

$$M_0 = (x_0, y_0, z_0), \quad M_1 = (x_1, y_0, z_1), \quad M_2 = (x_1, y_1, z_2), \quad M_4 = (x_0, y_1, z_3)$$

Consequently, notice that the point plotting sequence has an effect on this concept.

To display the surface, it must be divided into multiple facets whose data are stored inside three matrices (often denoted `xx, yy, zz` in the online help) that contain four lines and as many columns as facets. Each column in these matrices thus corresponds to one facet.

Caution › *Sometimes, two out of the four points of the facet are identical. In this case, the face is a triangle rather than a rectangle.*

In practice, you won't necessarily have to construct matrices to define a surface's facets. Several types of figure can arise, depending on the context:

The surface is a plot of a function of two variables $z = f(x, y)$

Calculating facets is not necessary. This computation will be directly performed by Scilab's graphics functions (such as `surf`) from the `surf(x,y,z)` command where:

- `x,y` are two vectors that define the grid (a discretization of the coordinate values)
- `z` is a matrix that satisfies `z(i,j)= f(x(i),y(j))` and is the height of the points in the grid defined by the vectors `x,y`

To compute the matrix `z`, there are two possible choices:

- Use the function `meshgrid` which, starting from two vectors `x,y`, generates the matrices `X,Y` for which `X(i,j)=x(i)` and `Y(i,j)=y(j)`. We can then combine these matrices to compute the height associated to the points in the grid as shown in the following example:

```
-->// grid definition

-->x=[0:2],y=x,
 x  =
    0.    1.    2.
 y  =
    0.    1.    2.

-->// surface calculation

-->[X,Y]=meshgrid(x,y)
 Y  =
    0.    0.    0.
    1.    1.    1.
    2.    2.    2.
 X  =
    0.    1.    2.
    0.    1.    2.
    0.    1.    2.

-->Z=X-Y
 Z  =
    0.    1.    2.
  - 1.    0.    1.
  - 2.  - 1.    0.

-->//display the surface

-->clf;

-->surf(x,y,Z)
```

- Define a function `z=f(x,y)` and use the command `feval` to calculate the values of `f` over the grid.

```
-->function z=plan(x,y)
-->    z=x-y
-->endfunction

-->// evaluate over a grid

-->x=[0:2],y=x,z=feval(x,y,plan)'
 x  =
    0.    1.    2.
 y  =
```

```
      0.      1.      2.
  z  =
      0.      1.      2.
    - 1.      0.      1.
    - 2.    - 1.      0.

  -->// display the surface

  -->clf;surf(x,y,z)
```

In both cases, you will get the surface in Figure 21.7.

Figure 21.7 : *Displaying a surface with* `surf` *for a plot a function of two variables*

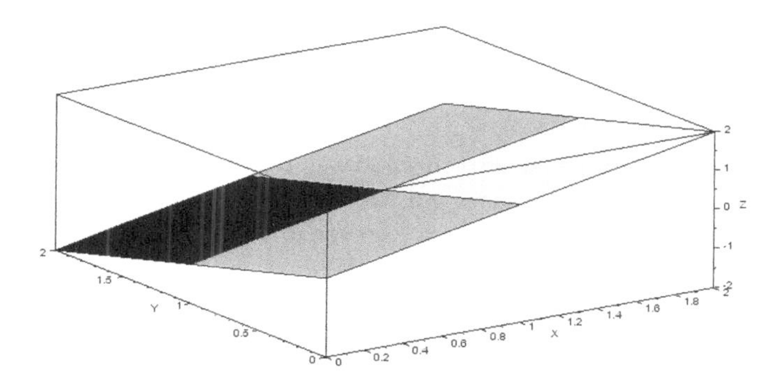

Caution › *In the example that uses* **`feval`** *do not forget to transpose the result given by* **`feval`**. *If the grid provided by the* `x, y` *vectors is not square but rectangular and you forget to transpose, it will result in an error stating* Wrong size for input argument 'Y'.

```
-->function z=plan(x,y)
-->    z=x-y
-->endfunction

-->// rectangular grid

-->x=[0:4],y=[0:2],
 x  =
    0.    1.    2.    3.    4.
 y  =
    0.    1.    2.

-->// bad evaluation

-->z=feval(x,y,plan)
 z  =
    0.  - 1.  - 2.
```

```
    1.    0.  - 1.
    2.    1.    0.
    3.    2.    1.
    4.    3.    2.

-->// surface is correctly displayed with plot3d

-->clf;plot3d(x,y,z)

-->// incompatible display dimensions

-->clf;surf(x,y,z)

    !--error 10000

 surf: Wrong size for input argument 'Y': A vector of size 5
 expected.
```

If you prefer to use **plot3d**, *the display conventions are reversed! There is therefore no need to transpose the result given by* **feval**, *however you will need to transpose the results obtained with* **meshgrid**.

If you need to compute the facets associated with a plot of a function of two variables, you will need to use the command `genfac3d` on the matrices `x`, `y`, `z` to generate matrices that define the facets. Then, you need to display them with a command such as `plot3d` or `plot3d1`:

```
-->function z=plan(x,y)
-->    z=x-y
-->endfunction

-->// evaluate over a grid

-->x=[0:2],y=x,z=feval(x,y,plan)
 x  =
    0.    1.    2.
 y  =
    0.    1.    2.
 z  =
    0.  - 1.  - 2.
    1.    0.  - 1.
    2.    1.    0.

-->// compute the facets

-->[xx,yy,zz]=genfac3d(x,y,z)
 zz  =
    0.    1.  - 1.    0.
  - 1.    0.  - 2.  - 1.
    0.    1.  - 1.    0.
    1.    2.    0.    1.
 yy  =
    0.    0.    1.    1.
```

```
    1.    1.    2.    2.
    1.    1.    2.    2.
    0.    0.    1.    1.
 xx  =
    0.    1.    0.    1.
    0.    1.    0.    1.
    1.    2.    1.    2.
    1.    2.    1.    2.

-->//display the facets

-->clf;plot3d1(xx,yy,zz)
```

Figure 21.8 : Computing facets with `genfac3d` for a plot of a function of two variables

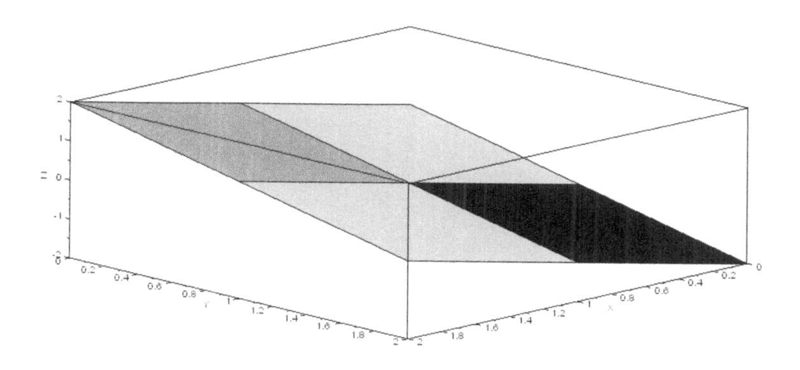

The surface is not a plot of a function of two variables

We then call it a parametric surface. You will have to calculate the faces by using the function `eval3dp` before being able to display the surface with `plot3d` or `plot3d1`:

```
-->function [x,y,z]=plan(u,v)
-->    x=u
-->    y=v
-->    z=u-v
-->endfunction

-->// compute parameters

-->u=[0:2],v=u,
 u  =
    0.    1.    2.
 v  =
    0.    1.    2.
```

```
-->// compute the facets

-->[xx,yy,zz]=eval3dp(plan,u,v)
 zz  =
    0.    1.  - 1.    0.
    1.    2.    0.    1.
    0.    1.  - 1.    0.
  - 1.    0.  - 2.  - 1.
 yy  =
    0.    0.    1.    1.
    0.    0.    1.    1.
    1.    1.    2.    2.
    1.    1.    2.    2.
 xx  =
    0.    1.    0.    1.
    1.    2.    1.    2.
    1.    2.    1.    2.
    0.    1.    0.    1.

-->//display the facets

-->clf;plot3d1(xx,yy,zz)
```

Figure 21.9 : *Computing facets with* `eval3dp` *for a parametric surface*

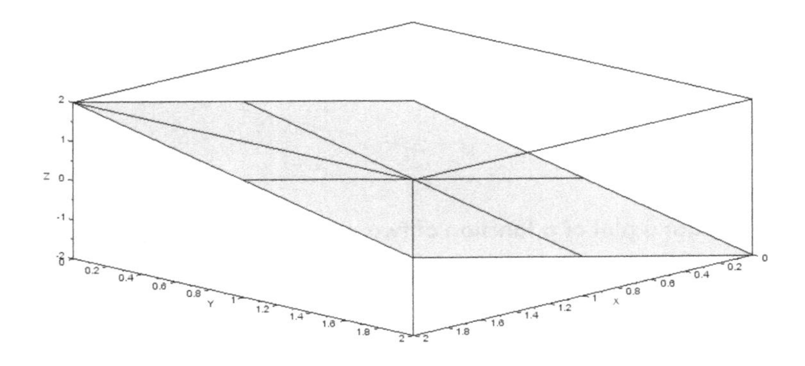

For surface displays, you can ask Scilab to color the facets that make up the surface. You have two options for coloring at your disposal:

- All the facets are the same color as shown in Figure 21.9.
- The facet colors change as a function of the points `(x(i),y(j),z(i,j))` which define the surface as in Figure 21.7 and Figure 21.8.

For this last case, depending on the graphics function used, Scilab defines the values `C(i,j)` associated to each point in the plot then uses a color table to match a color to these values. A color table is in fact a matrix with three columns where each row defines a color's RGB levels. This color table is stored inside the `color_map` property of the Figure handle linked to the plot.

```
-->lines(10);          // limits the number of rows displayed

-->F=gcf();            // new figure

-->size(F.color_map)  // table of 32 colors
 ans  =
    32.    3.

-->F.color_map         // default table of colors
 ans  =
    0.          0.          0.
    0.          0.          1.
    0.          1.          0.
    0.          1.          1.
    1.          0.          0.
```

Several functions let you create color gradients to color surfaces according to your own tastes! For example, execute the following script to get the plots in Figure 21.10.

```
F=scf(1);clf;
plot3d()
F.color_map=hotcolormap(64);
xtitle('hotcolormap')
F=scf(2);clf;
plot3d()
F.color_map=jetcolormap(64);
xtitle('jetcolormap')
F=scf(3);clf;
plot3d()
F.color_map=oceancolormap(64);
xtitle('oceancolormap')
F=scf(4);clf;
plot3d()
F.color_map=graycolormap(64);
xtitle('graytcolormap')
```

Here, we get a table of 64 colors with the help of different Scilab functions:

- `hotcolormap` for a brown-red-yellow-white gradient
- `jetcolormap` for a blue-green-yellow-red gradient
- `graycolormap` for a gray gradient
- `oceancolormap` for a blue gradient

You can find other options in the online help (type `help colormap`).

Figure 21.10 : *Different color tables for surface plots*

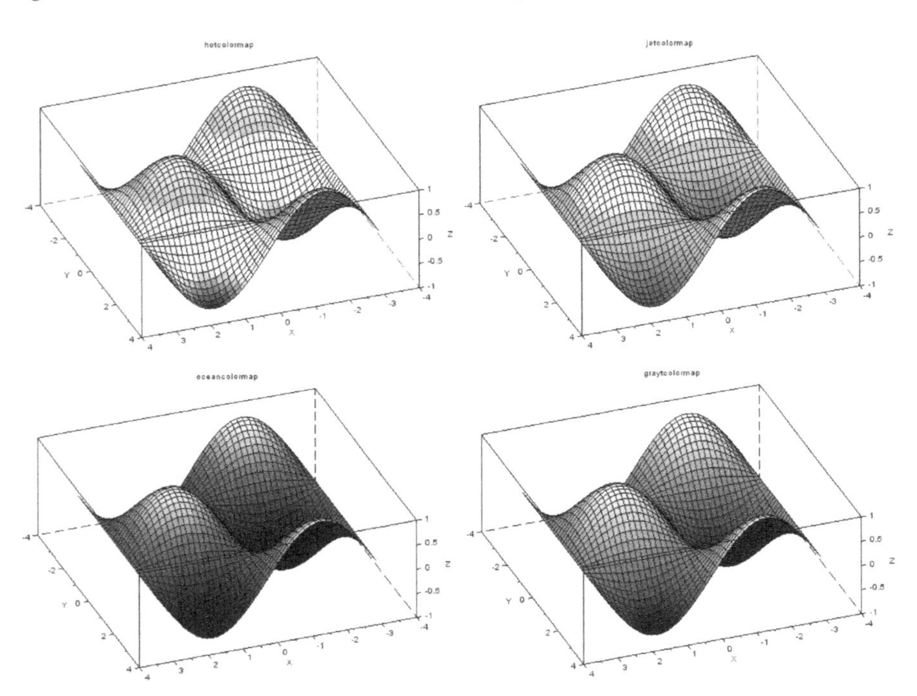

Caution › *Certain functions use the color table shown in* ***Figure*** *20.10 by default, although it is not very appealing. Note that there can only be one color table per Figure handle. If a function contains several Axes, these must use the same color table.*

To better visualize the link between a color and the level of the `z` coordinate, you can add a color scale with `colorbar`:

```
// define a grid
x=[-1:0.1:1];y=x;
// surface computation
[X,Y]=meshgrid(x,y);
Z=X.^2-Y.^2;
// surface display
clf;
F=gcf();F.color_map=jetcolormap(64);
surf(x,y,Z)
colorbar(min(Z),max(Z))
```

In order to correctly set the color scale gradient as shown in Figure 21.11, you need to specify the extreme values of `min(z)` and `max(z)` in the arguments of the `colorbar` function. The list of colors is taken from the color table associated to the Figure handle.

Figure 21.11 : *Color scale*

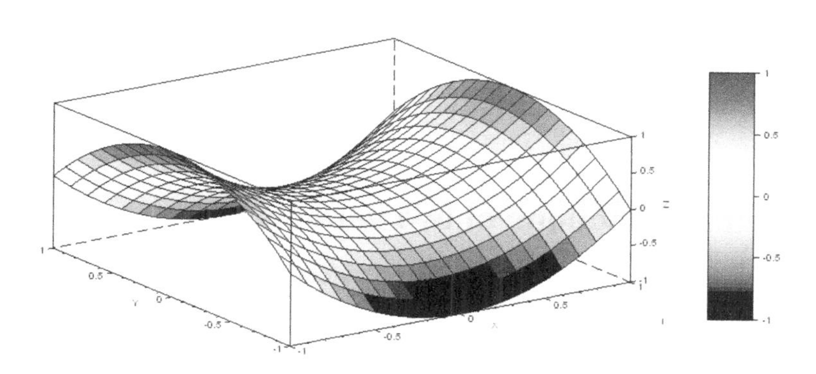

Tip › *You can plot surfaces without displaying certain facets. To do this, you need to substitute the value* `z(i,j)` *with* **%nan** *for points that you wish to omit from the facet list. You will then get a "gap" in the surface corresponding to the facets that are not displayed. In the following script, the values of the square* $-0.4 \le x \le 0.4$ *and* $-0.4 \le y \le 0.4$ *were replaced with* **%nan**. *This yields the figure* **below**.

```
// define a grid
x=[-2:0.2:2];y=x;
// surface computation
[X,Y]=meshgrid(x,y);Z=X-Y;
// Z=%nan for a square in the center
indices=find((abs(X)<0.4)&(abs(Y)<0.4));
Z(indices)=%nan;
// surface display
clf;
plot3d(x,y,Z)
```

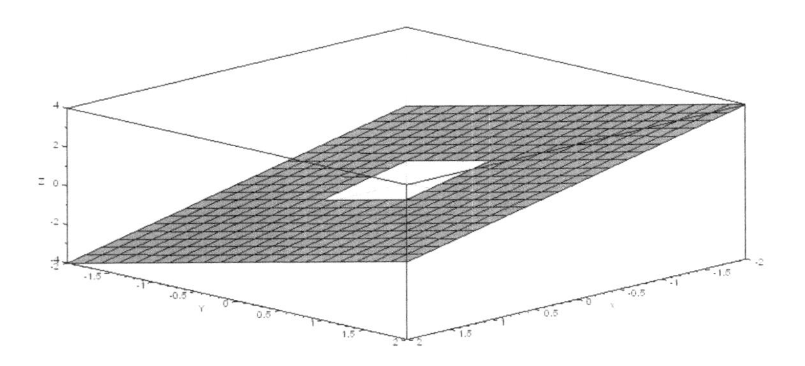

21.4. Plotting functions of two variables

We are now going to study several examples of surfaces to figure out which type of command is better suited to a given display. At the same time, you will have the opportunity to study the handle properties of Fac3d and Plot3d that are associated to the surfaces. The main commands used to display surfaces are:

- `surf` and `mesh` whose syntax will be familiar for MATLAB users
- `plot3d` and `plot3d1` which accept numerous options (specific to Scilab) to modify the surface rendering

To get started, we will plot a saddle surface by using these four graphics functions and compare their results. This surface is a plot of a function of two variables. After executing the following script, you will get the four plots in Figure 21.12.

```
function z=saddle(x,y)
    z=x^2-y^2
endfunction
// evaluate over a grid
x=[-1:0.2:1];y=x;z=feval(x,y,saddle);
// surface display
clf;
subplot(221)
surf(x,y,z')
xtitle('surf')
subplot(222)
mesh(x,y,z')
xtitle('mesh')
subplot(223)
plot3d(x,y,z)
xtitle('plot3d')
subplot(224)
```

```
plot3d1(x,y,z)
xtitle('plot3d1')
```

Figure 21.12 : *Different renderings for surfaces displayed with* `surf`*,* `mesh`*,* `plot3d` *and* `plot3d1`

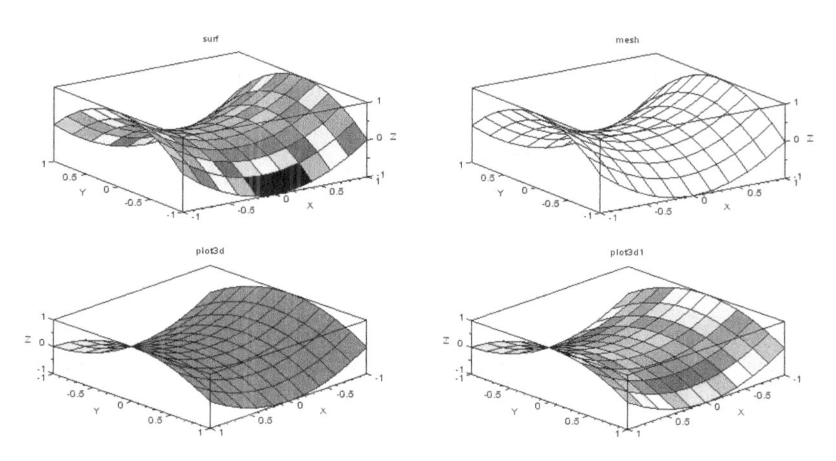

You will notice that for identical data, you get different types of displays! This is due to the fact that each command has its own way of processing the handle properties associated to the different plots. Moreover, the handles of the four plots are different:

- Fac3d for plots obtained with `surf` and `mesh`
- Plot3d for plots obtained with `plot3d` and `plot3d1`

You can check this by comparing the values of `F.children.children` et `F.children.title.text` where `F` is the figure handle:

```
-->F=gcf();  // figure handle

-->// the four surfaces: 2 Plot3d and 2  Fac3d

-->F.children.children
 ans  =
1 by 4 matrix of handles:
=========================

-->// name of each plot
```

```
-->F.children.title.text
 ans  =
!plot3d1  plot3d  mesh  surf  !
```

Nevertheless, these handles all define surfaces and have numerous properties in common. This is also the case for the Axes handles that hold the plots. You can see from the example in Figure 21.12 that the view angle varies according to the function used.

```
-->// compairing view angles

-->R=zeros(2,4);R(:)=F.children.rotation_angles
 R  =
    35.     35.     51.      51.
    45.     45.   - 125.   - 125.
```

- `alpha=51` and `theta=-125` for `surf` and `mesh`
- `alpha=35` and `theta=45` for `plot3d` and `plot3d1`

To understand the different appearance of the four surfaces, you have to compare the property values of the Axes, Fac3d and Plot3d axes that are associated to the plots. For example, to compare the colors of the surfaces, look at the values of `color_flag`, `color_mode` and `hiddencolor`.

```
-->// display colors

-->F.children.children.color_flag
 ans  =
    1.    0.    0.    4.

-->F.children.children.color_mode
 ans  =
    2.    2.    8.    2.

-->F.children.children.hiddencolor
 ans  =
    4.    4.    0.    0.
```

You will notice that:

- The value of the `color_flag` property determines the way in which the *visible* face of a surface's facets get colored:
  - with a uniform color when the value is 0 (this is the case with `mesh` and `plot3d`)
  - by using the Figure handle's color table for all other values (for example with `surf` and `plot3d1`)

- When the surface is colored uniformly, `color_mode` determines which color is used (see the color table in Figure 20.10) :
  - For the function `mesh`, this value is 8 which corresponds to white in the *default* color table
  - For the function `plot3d`, this value is 2 which corresponds to blue in the *default* color table
- Finally, the `hiddencolor` property determines which color gets used for the lower face of the facets:
  - When this property is assigned the value 0, the lower face of each facet adopts the same color as the upper face. This is the case when using `surf`.
  - When this property has a different value, the lower face of the facets is colored uniformly; for example, in cyan for the value 4 (see the color table in Figure 20.10). This is the case when using `plot3d` or `plot3d1`.

Additionally, you can change the plot's appearance just by modifying certain properties. For example, you can easily isolate or remove facets from a plot as shown in the following script which creates Figure 21.13.

```
// evaluate over a grid
x=[-1:0.2:1];y=x;[X,Y]=meshgrid(x,y);
Z=X.^2-Y.^2;
// change the color table
clf;
F=gcf();F.color_map=jetcolormap(8);
// surface display
subplot(221)
surf(x,y,Z)
E=gce();E.color_mode=0;
xtitle('color_mode=0')
subplot(222)
surf(x,y,Z)
E=gce();E.color_mode=1;
xtitle('color_mode=1 (by default)')
subplot(223)
surf(x,y,Z)
E=gce();E.color_mode=-1;
xtitle('color_mode=-1')
subplot(224)
surf(x,y,Z)
E=gce();E.color_mode=4;E.color_flag=0;
xtitle('color_mode=4, color_flag=0')
```

Figure 21.13 : *Plotting with or without facets as a function of the* `color_mode` *property*

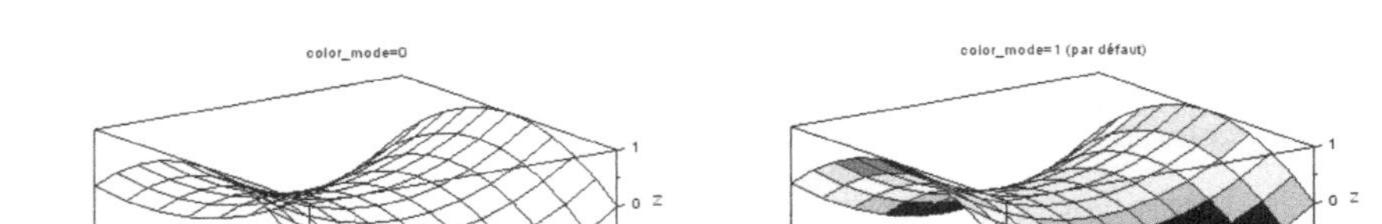

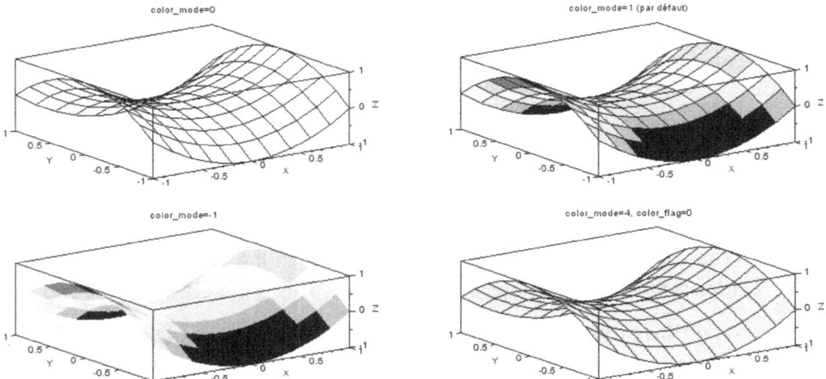

The `surf` and `mesh` commands let you more easily associate a color to the facets or modify the axes tick marks:

- `surf(x,y,z)` or `mesh(x,y,z)` plot facets associated to the points `(x(i),y(j),z(i,j))`, the axes `x,y` have tick marks located as a function of the values of `x,y`.
- `surf(z)` or `mesh(z)` plot facets associated to the points `(i,j,z(i,j))`, such that the `x,y` axes have tick marks as a function of the values of the indices `i,j`.
- `surf(x,y,z,C)` or `surf(z,C)` plot facets colored with the colors `C(i,j)` (where `C` is the same size as `z`).

```
// evaluate over a grid
x=[-1:0.2:1];y=x;[X,Y]=meshgrid(x,y);
Z=X.^2-Y.^2;
// compute a color (0,1,...,10) for each facet
C=round(5*(1+X));
// display the surface
clf;
subplot(221)
surf(Z)
xtitle('surf(z)')
subplot(222)
surf(Z,C)
xtitle('surf(z,C)')
subplot(223)
surf(x,y,Z,C)
```

```
xtitle('surf(x,y,z,C)')
subplot(224)
mesh(Z)
xtitle('mesh(z)')
```

Figure 21.14 : Different uses of `surf`

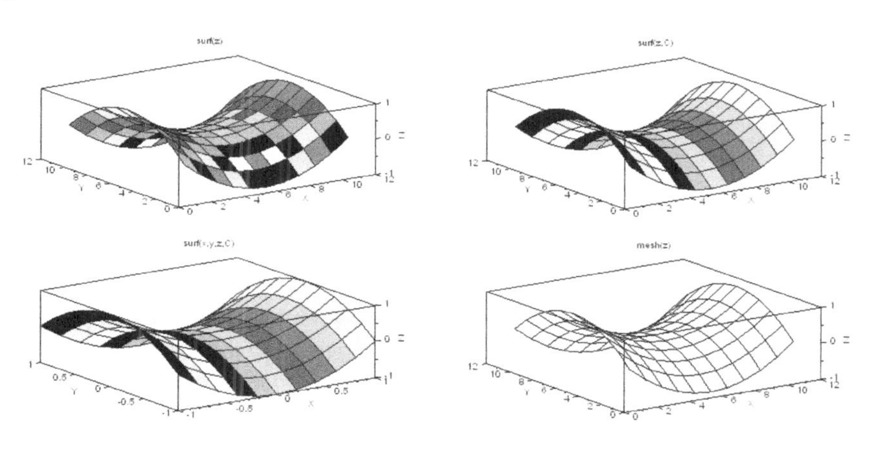

Tip › *When you wish to plot a Scilab function of two variables, you can use* **`fplot3d`** *or* **`fplot3d1`** *which lets you avoid having to compute the values of the function to plot on the grid defined by the* `x, y` *vectors.*

```
function z=saddle(x,y)
    z=x^2-y^2
endfunction
// define the grid
x=[-1:0.2:1];y=x;
// display the surface
clf;
subplot(121)
fplot3d(x,y,saddle)
xtitle('fplot3d')
subplot(122)
fplot3d1(x,y,saddle)
xtitle('fplot3d1')
// color table
F=gcf();F.color_map=jetcolormap(8);
```

The renderings of the plots shown **below** *are those associated with what would be obtained with* **`plot3d`** *or* **`plot3d1`** *by evaluating the* **`saddle`** *function over a grid defined by the* `x, y` *vectors.*

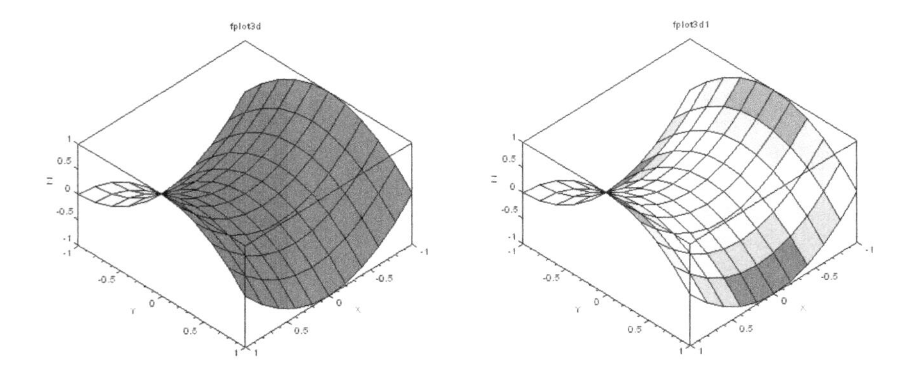

21.5. Parametric surfaces

Displaying parametric surfaces is done in much the same way as graphing functions of two variables, even if the matrix provided as argument to the functions `surf`, `mesh`, `plot3d` or `plot3d1` now directly represents the surface's facets. Take the example of a torus. Execute the following script to get the plots in Figure 21.15.

```
function [x,y,z]=torus(theta,phi)
        R=1,r=0.2
        x=(R+r*cos(phi)).*cos(theta)
        y=(R+r*cos(phi)).*sin(theta)
        z=r*sin(phi)
endfunction
// compute facets
phi=[0:0.2:2*3.2];
theta=[2*3.2:-0.2:0];
[x,y,z]=eval3dp(torus,theta,phi);
// surface display
clf;
subplot(221)
surf(x,y,z)
xtitle('surf')
subplot(222)
mesh(x,y,z)
xtitle('mesh')
subplot(223)
plot3d(x,y,z)
xtitle('plot3d')
subplot(224)
plot3d1(x,y,z)
xtitle('plot3d1')
F=gcf();
F.color_map=hotcolormap(32);
// plot color of plot3d
```

```
F.children(2).children.color_mode=10;
// change the view angle of the 4 surfaces
F.children(1:2).rotation_angles=[80 135];
F.children(3:4).rotation_angles=[30 135];
```

You will find the same properties for these surfaces as for plots of functions of two variables according to the plotting command used.

Figure 21.15 : *Plotting parametric surfaces with* `surf`, `mesh`, `plot3d` *and* `plot3d1`

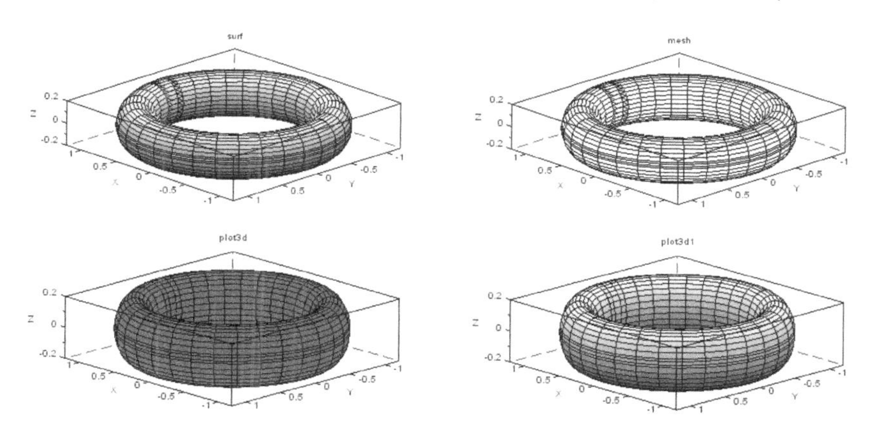

Caution › *In order to correctly compute facets with the* **`eval3dp`** *function, it is very important that the function that defines the parametric surface is coded with element-wise operations* `.*, .^, ./, \.` *rather than the standard operations* `*, ^, /, \`*. Indeed the latter can be mistaken for the corresponding matrix operations when computing facets. This is the reason the* `.*` *operation is used in the* **`torus`** *function's script rather than* `*` *as shown in* ***Figure 21.15****.*

Several commands are available to deal with parametric surfaces:

- `plot3d2` plots a surface with uniform colors, like `plot3d`, except the purpose of the upper and lower faces is reversed.
- `plot3d3` plots grids associated to the facets of a parametric surface, similar to `mesh`.

You can compare the renderings of the different commands `plot3d*` in Figure 21.16 which was obtained by executing the following script:

```
function [x,y,z]=torus(theta,phi)
        R=1,r=0.2
        x=(R+r*cos(phi)).*cos(theta)
        y=(R+r*cos(phi)).*sin(theta)
        z=r*sin(phi)
```

```
endfunction
// compute facets
phi=[0:0.1:2*3.15];
theta=[2*3.16:-0.1:0];
[x,y,z]=eval3dp(torus,theta,phi);
// surface display
clf;
subplot(221)
plot3d(x,y,z)
xtitle('plot3d')
subplot(222)
plot3d1(x,y,z)
xtitle('plot3d1')
subplot(223)
plot3d2(x,y,z)
xtitle('plot3d2')
subplot(224)
plot3d3(x,y,z)
xtitle('plot3d3')
// change the view angle of the 4 surfaces
F=gcf();F.children.rotation_angles=[80 45];
```

Figure 21.16 : *Different renderings for parametric surface displays with* `plot3d`*,* `plot3d1`*,* `plot3d2` *and* `plot3d3`

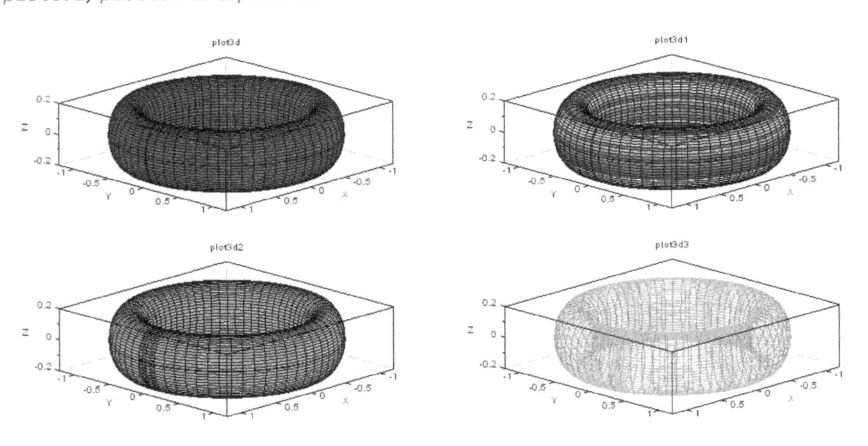

Despite appearances, the `plot3d3` command constructs a more complex grid than `mesh`. It involves the superposition of two grids made up of *polyline* type entities. You can execute the following script to separate the two grids (see Figure 21.17).

```
function [x,y,z]=torus(theta,phi)
        R=1,r=0.2
        x=(R+r*cos(phi)).*cos(theta)
        y=(R+r*cos(phi)).*sin(theta)
```

```
        z=r*sin(phi)
endfunction
// compute facets
phi=[0:0.1:2*3.15];
theta=[2*3.16:-0.1:0];
[x,y,z]=eval3dp(torus,theta,phi);
// surface display
clf;
subplot(121)
plot3d3(x,y,z)
A=gca();delete(A.children(2))
xtitle('A=gca();A.children(1)')
subplot(122)
plot3d3(x,y,z)
A=gca();delete(A.children(1))
xtitle('A=gca();A.children(2)')
// change the view angle of the 4 surfaces
F=gcf();F.children.rotation_angles=[80 45];
```

The first grid is made up of horizontal lines (in cyan) and the second of horizontal and vertical lines (in green).

Figure 21.17 : *Grid of a surface created with* `plot3d3`

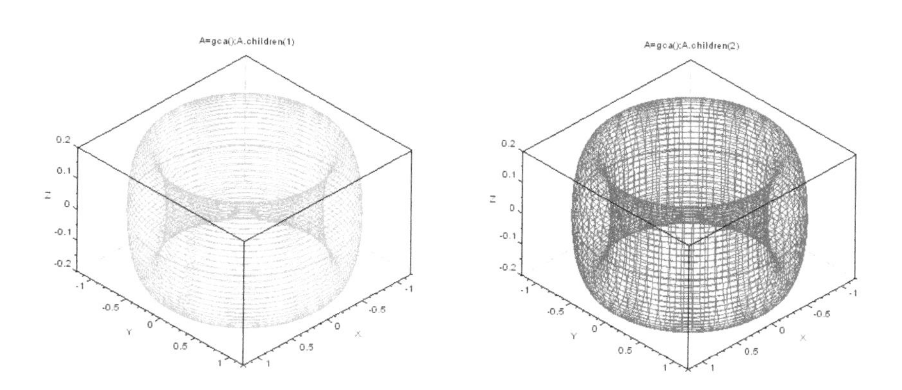

Caution › *Even if viewed from the same observation point, figures plotted with* **`plot3d`** *or* **`plot3d1`** *may appear different from those obtained with* **`surf`** *or* **`mesh`**.

```
function [x,y,z]=torus(theta,phi)
        R=1,r=0.2
        x=(R+r*cos(phi)).*cos(theta)
        y=(R+r*cos(phi)).*sin(theta)
        z=r*sin(phi)
endfunction
```

```
// compute facets
phi=[0:0.1:2*3.15];
theta=[2*3.16:-0.1:0];
[x,y,z]=eval3dp(tore,theta,phi);
// surface display
clf;
subplot(121)
surf(x,y,z)
xtitle('surf')
subplot(122)
plot3d1(x,y,z)
xtitle('plot3d')
// same view angles
F=gcf();
F.children.rotation_angles=[80 100];
// change the color table
F.color_map=hotcolormap(64)
```

You can see in the figure **below** *that the observation point corresponds to the same* **`rotation_angles`** *values but that the rendering is different.*

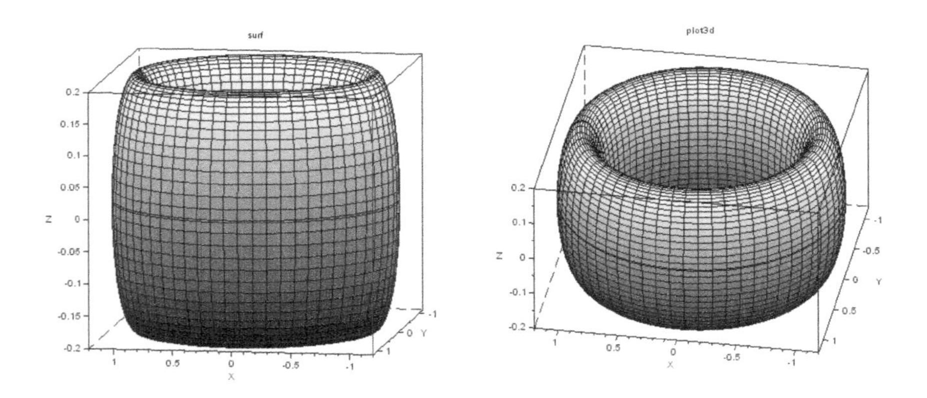

With the `plot3d` command, several options are available, following the format `key=value`, to easily modify the properties of the current Axes handle (or its children) without having to call `gda()`. For example, you can:

- set the observation direction with the options `alpha`, `theta`, which modify the `rotation_angles` property
- specify the region's bounds with the `ebox` option, with the format `ebox=[xmin,xmax,ymin,ymax,zmin,zmax]` or `ebox=[xmin,ymin,zmin;xmax,ymax,zmax]`, which will modify the `data_bounds` property (see section *Zooming*)

- change the labels of the coordinate axes with the `leg` option by following the format `leg="X@Y@Z"` : the strings that substitute `X,Y,Z` will replace the values of the `x_label`, `y_label`, `z_label` properties
- change the property values of `color_mode`, `axes_visible`, etc. by using the option `flag`, with the syntax `leg=[a,b,c]`

The following script provides a demonstration.

```
// evaluate over a grid
x=[-1:0.2:1];y=x;
[X,Y]=meshgrid(x,y);
Z=X.^2-Y.^2;
// surface display
clf;
plot3d(x,y,Z,alpha=50,theta=140,leg="X1@X2@X3",flag=[0 0 4])
A=gca();
// values of alpha and theta
A.rotation_angles
// axes labels
[A.x_label.text,A.y_label.text,A.z_label.text]
// first argument for flag
A.children.color_mode
```

The Figure 21.18 gets returned after the script execution and accurately displays the plot with `color_mode=0` and the axes labels are changed to `X1,X2,X3`.

Figure 21.18 : *Examples of Options with* `plot3d`

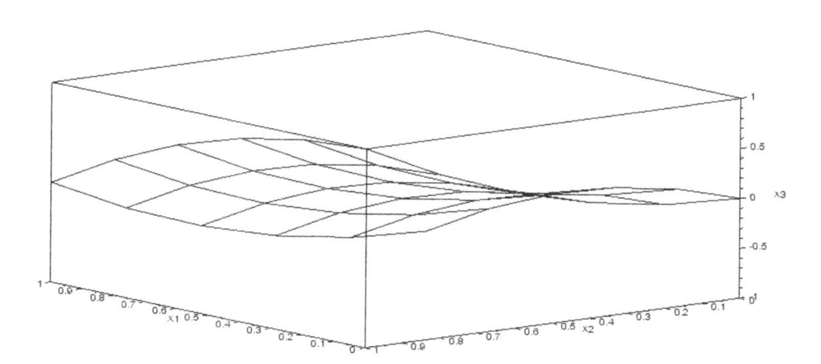

You can easily relate this to the figure's handle values which are displayed in the console:

```
-->// evaluate over a grid

-->x=[-1:0.2:1];y=x;

-->[X,Y]=meshgrid(x,y);

-->Z=X.^2-Y.^2;

-->// surface display

-->clf;

-->plot3d(x,y,Z,alpha=50,theta=140,leg="X1@X2@X3",flag=[0 0 4])

-->A=gca();

-->// values of alpha and theta

-->A.rotation_angles
 ans  =
    50.    140.

-->// axes labels

-->[A.x_label.text,A.y_label.text,A.z_label.text]
 ans  =
!X1  X2  X3  !

-->// first argument for flag

-->A.children.color_mode
 ans  =
    0.
```

Caution › The syntax for the `plot3d` *options, although a little complex, is compatible with numerous other commands like* `plot3d*`.

21.6. Representation of 2D surfaces

The `grayplot` and `Sgrayplot` commands let you display 2D surfaces.

```
// grid definition
x=[-1:0.1:1];y=x;
// surface calculation
[X,Y]=meshgrid(x,y);
Z=X.^2-Y.^2;
// display the surface
clf;
F=gcf();F.color_map=hotcolormap(64);
subplot(121)
grayplot(x,y,Z)
xtitle('grayplot')
```

```
subplot(122)
Sgrayplot(x,y,Z)
xtitle('Sgrayplot')
```

These two functions take the same arguments as input as functions of the `plot3d` type. The two plot's differences in rendering in Figure 21.19 stem from the way in which the facets are colored as a function of the values of the four corners.

- `grayplot` colors each facet with one color as a function of the average of the four corners.
- `Sgrayplot` colors each facet with a color gradient computed from the interpolation of the four corners' values. This creates a smoother transition between facets.

Figure 21.19 : *Surfaces color levels*

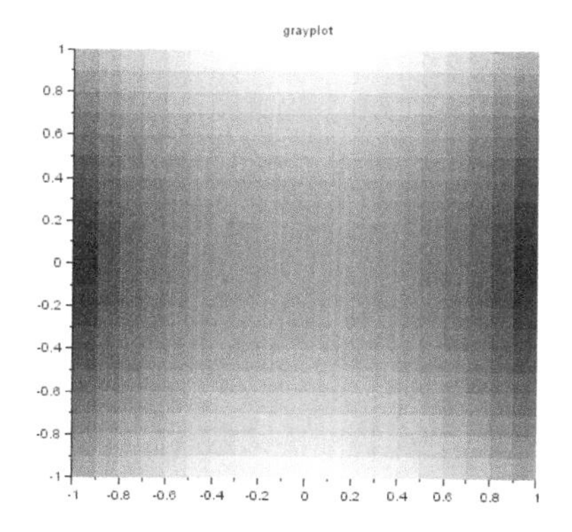

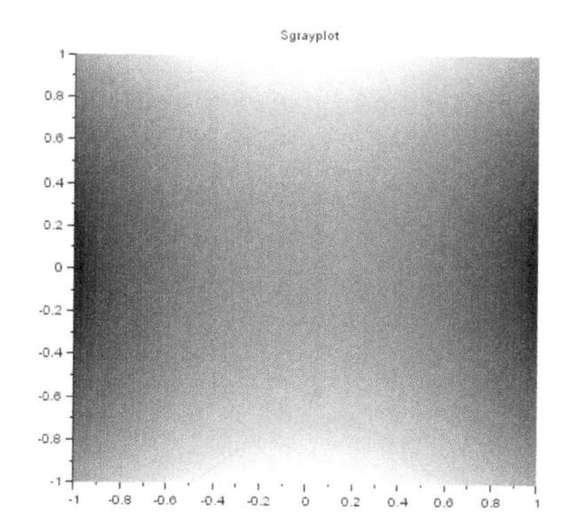

As with `plot3d`, we have the two commands `fgrayplot` and `Sfgrayplot`, which are similar to `grayplot` and `Sgrayplot` but avoid having to evaluate the function to plot. The 2D surface plots correspond to handles of type *Fec* and *Grayplot*, depending on whether the surface coloring scheme is interpolated or not. You can check this by looking at the handle in Figure 21.19.

```
-->F=gcf();

-->F.children.title.text
 ans  =
!Sgrayplot  grayplot  !

-->// Sgrayplot handle

-->F.children.children(1).children.type
```

```
ans  =
Fec

-->// grayplot handle

-->F.children.children(2).type
ans  =
Grayplot
```

Caution › If you use a color table with 256 shades of gray, created with **`graycolormap`**, *then the surfaces displayed by* **`grayplot`** *and* **`Sgrayplot`** *will be grayscale images for which the shades are stored with 8 bits. Execute the following script to generate the figure* **below**.

```
x=-2:0.01:2; y=x;
function z=milkdrop(x,y)
    sq=(x^2+y^2)
    z= exp( exp(-
sq).*(exp(cos(sq).^20)+8*sin(sq).^20+2*sin(2*(sq)).^8) );
endfunction
z=feval(x,y,milkdrop)';
clf;
F=gcf();F.color_map=graycolormap(64);
subplot(121)
surf(x,y,z)
F.children(1).children.color_mode=-1;
xtitle('surf')
subplot(122)
Sfgrayplot(x,y,milkdrop)
xtitle('fgrayplot')
```

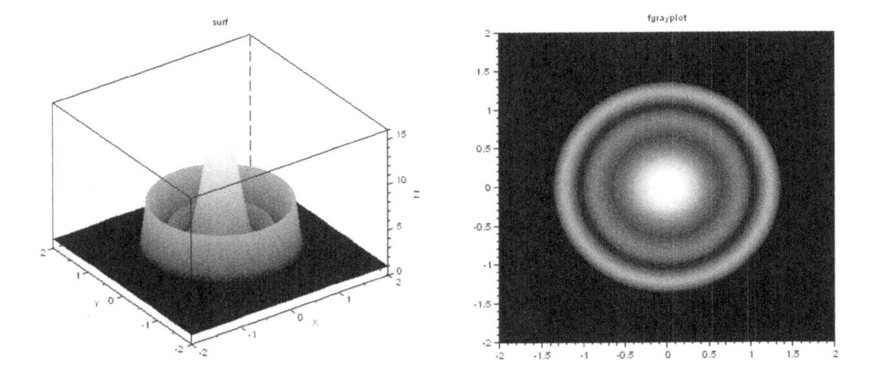

Scilab can also plot a curve composed of contour lines of different levels with the function `contour`. Execute the following script to get Figure 21.20.

```
// grid definition
x=[-1:0.1:1];y=x;
// surface computation
[X,Y]=meshgrid(x,y);
Z=X.^2-Y.^2;
// surface display
clf;
subplot(121)
contour(x,y,Z,7)
xtitle('contour(x,y,Z,7)')
subplot(122)
L=[-0.6 -0.3 -0.1 0 0.2 0.5 0.7];
contour(x,y,Z,L)
xtitle('contour(x,y,Z,L)')
```

The first three arguments of the contour function define the surface and the fourth indicates which contours to plot. You can let Scilab choose the line contour levels by providing the number of levels to plot (first example) or by providing a list of levels (second example). Here again, the colors are chosen from the current figure's color table.

Figure 21.20 : *Plotting contour lines with* contour

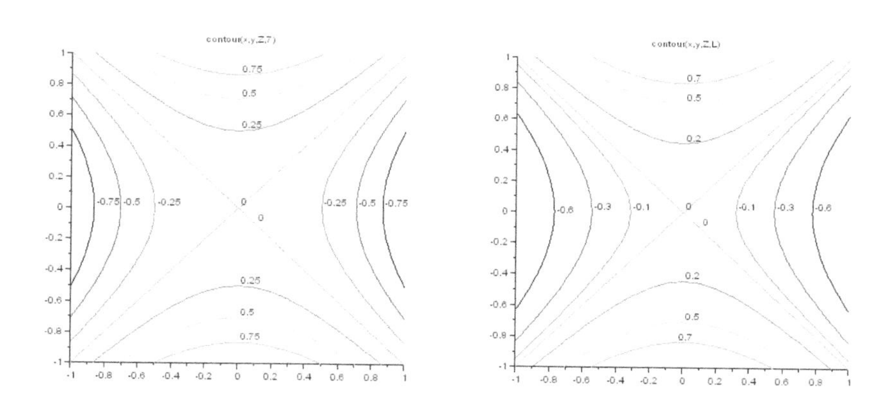

You can also superimpose contour lines to a plot that was previously generated by using grayplot or Sgrayplot or even add it to a surface plotted in three dimensions with plot3d1 or plot3d.

```
// grid definition
x=[-1:0.1:1];y=x;
// surface computation
[X,Y]=meshgrid(x,y);
```

```
Z=X.^2-Y.^2;
// surface display
clf;
F=gcf();F.color_map=hotcolormap(64);
subplot(121)
// plots a surface in 3D
plot3d1(x,y,Z)
// remove the facet plot
E=gce();E.color_mode=-1;
// add the contour lines to the surface
contour(x,y,Z,7,flag=[0 2 4])
xtitle('plot3d1+contour')
subplot(122)
// plot the surface in 2D
Sgrayplot(x,y,Z)
// add the contour lines to the surface
contour(x,y,Z,7)
xtitle('Sgrayplot+contour')
```

The `0` value in the `flag` option that gets passed to the `contour` function is responsible for adjusting the contour lines to the 3D curve. You can rotate the plot in Figure 21.21 to ascertain that the contour lines are indeed stuck to the surface.

***Figure 21.21** : Superimposing contour curves to a surface*

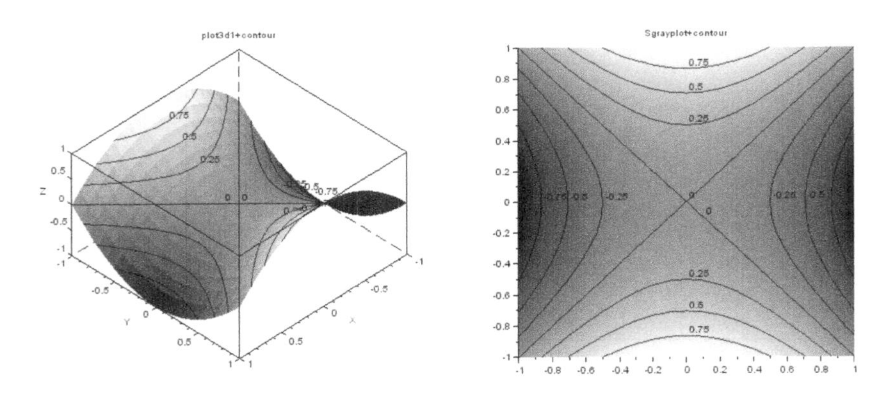

***Caution** › It is impossible to superimpose contour lines to a plot created using* **`surf`**. *In this case, you need to use* **`plot3d1`**.

If you'd prefer to fill the interval between two contour lines, privilege the use of the `contourf` command, as shown in the following script.

```
// grid definition
x=[-1:0.1:1];y=x;
```

```
// surface computation
[X,Y]=meshgrid(x,y);
Z=X.^2-Y.^2;
// surface display
clf;
F=gcf();F.color_map=jetcolormap(8);
contourf(x,y,Z,7)
xtitle('contourf(x,y,Z,7)')
```

You should get the plot in Figure 21.22. Here we have chosen to use a table of eight colors so that each of the seven levels has its own assigned color.

Figure 21.22 : *Filling contour curves with* `contourf`

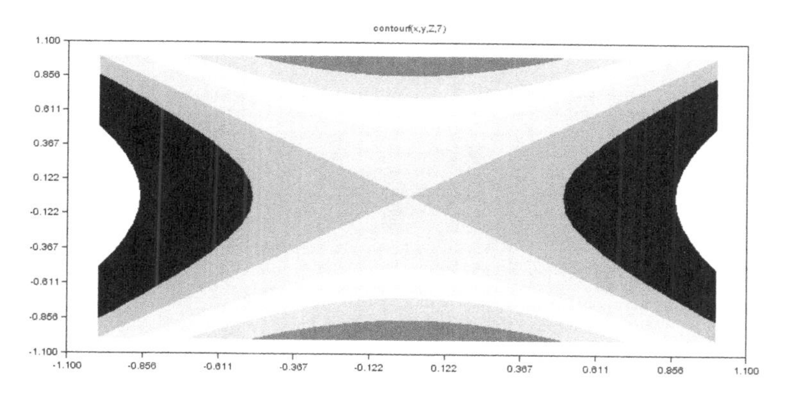

The functions we have seen so far color each facet by using the values of its four corners. To color a facet with a well-defined value, you will need to use `Matplot` or `Matplot1`, which create handles of type `Matplot`. The following script constructs a matrix whose data are used to color the rectangles visible in Figure 21.23.

```
// grid definition
x=[1:5];y=[2:5];
// surface computation
[X,Y]=meshgrid(x,y);
Z=X+Y;
// surface display
clf;
F=gcf();F.color_map=jetcolormap(10);
subplot(121)
Matplot(Z)
xtitle('Matplot(Z)')
subplot(122)
A=gca();A.data_bounds=[0,0;10,10];
A.axes_visible=["on" "on" "on"];
Matplot1(Z,[1,2,5,5])
```

```
xtitle('Matplot1(Z,[1,2,5,5])')
```

In addition to the matrix used for coloring, the `Matplot1` function needs an argument that defines the rectangle in which the plot gets created. This argument is of the form `[xmin,ymin,xmax,ymax]`, as seen with the `rect` option in `plot2d`. The `Matplot` command on the other hand only asks for one mandatory input, but can accept optional arguments of the same type as `plot2d`.

Figure 21.23 : *Displaying matrices with colors with* `Matplot`

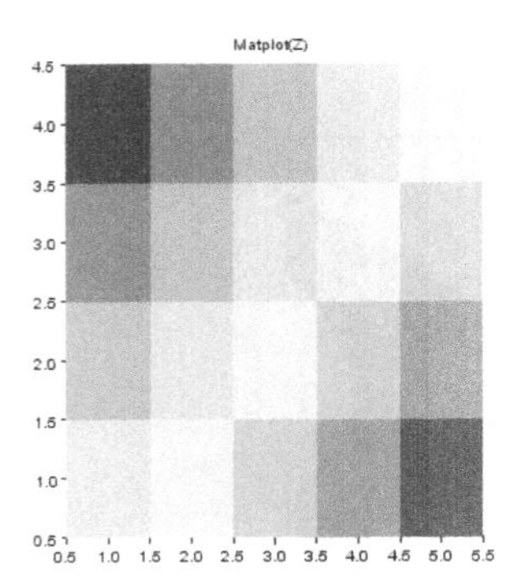

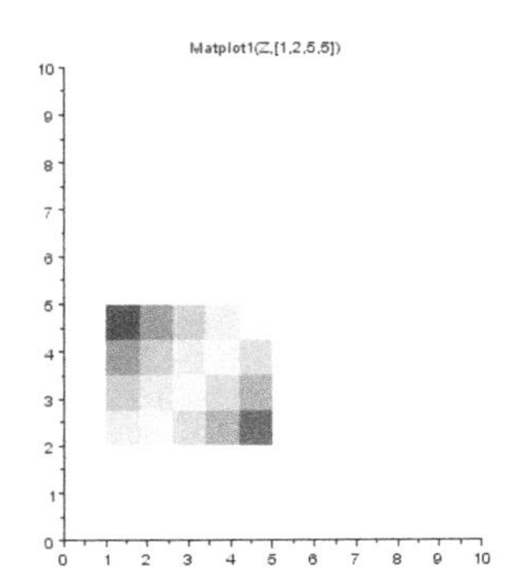

22

Other Two-dimensional Geometrical Elements

Scilab has specific commands that can be used to draw common geometrical shapes in a plane. In order to better understand how these different functions work, we will consistently resize the graphics window to the same size and include a visible isometric scale by adding a grid with the help of the commands:

```
clf;
plot2d(0,0,0,rect=[0,0,10,10],frameflag=3)
xgrid(4)  // grid
```

or even:

```
clf;
A=gca();A.data_bounds=[0 0 10 10];  // window size
A.axes_visible=["on" "on" "off"];  // make axes visible
A.isoview="on"   // isometric scale
xgrid(4)  // grid
```

***Caution** › Unless otherwise specified, the colors used come from the default color table and will often be denoted by their numbers (see **Figure 20.10**) rather than their names in order to simplify commands.*

22.1. Rectangles

A rectangle can be defined by four real numbers `x,y,w,h` where:

- `x,y` provide the coordinates of the upper-left corner of the rectangle.
- `w,h` provide the dimensions (length and height) of the rectangle.

To draw rectangles, use the functions `xrect` or `xfrect`, which take these four parameters as inputs. These two functions differ due to the fact that the `xrect` function only draws the contour of the rectangle while `xfrect` fills the rectangle with the color indicated by its name or number (see the table in Figure 20.10 only valid for the default color table):

```
clf;
plot2d(0,0,0,rect=[0,0,10,10],frameflag=3)
xgrid(4)  // grid
xfrect(0,4,2,4)  // first rectangle
E=gce(); E.background=1;  // black inside
xrect(2,4,2,2)   // second rectangle (square)
E=gce(); E.foreground=5;  // red outline
xrect(4,6,2,2)   // third rectangle (square)
E=gce();
E.foreground=2;  // blue outline
E.line_style=3;  // dotted line
E.thickness=5;  // thickness
xfrect(6,8,4,2)  // fourth rectangle
E=gce(); E.background=7;  // yellow inside
```

In fact, the two functions `xrect` and `xfrect` create a handle of type *rectangle*. The only difference lies in this handle's `fill_mode` property, which is set to `"off"` for the contour and `"on"` for a filled rectangle. Depending on whether the outline is colored or the inside, the color number used will be inside the `foreground` or `background` property.

Figure 22.1 : *Drawing rectangles with `xrect` and `xfrect`*

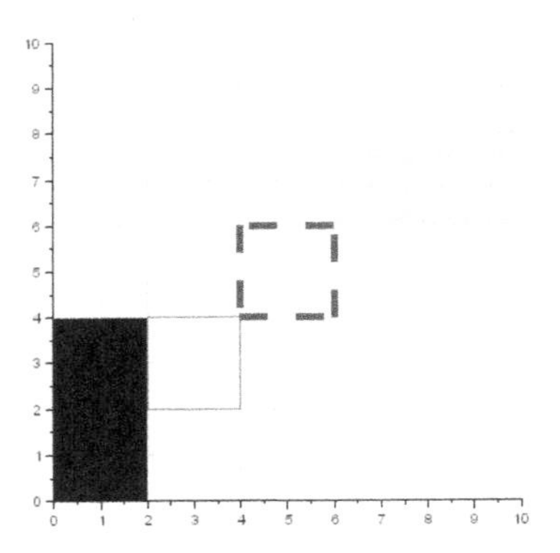

Tip › You can call the commands `xrect` *and* `xfrect` *by passing arguments formatted as one vector with four components, meaning* `xrect([x,y,w,h])` *instead of* `xrect(x,y,w,h)`.

In order to plot several rectangles with just one command, use `xrects`. This function takes two arguments as inputs:

- one matrix where each column corresponds to the parameters `x,y,w,h` of a rectangle
- a `fill` vector containing the numbers corresponding to the colors to use:

- `fill(i)>0` to color the inside of the `i`th rectangle with the color `fill(i)`
- `fill(i)<0` to color the outline of the `i`th rectangle with the color `-fill(i)`

You can execute the following script to get the result in Figure 22.2.

```
clf;
plot2d(0,0,0,rect=[0,0,10,10],frameflag=3)
xgrid(4)
// tracing "full" rectangles
rects1=[
2,2,2,2;
4,4,2,2;
6,6,2,2
8,8,2,2]';
xrects(rects1,[2:5])
// plotting rectangle contours
rects2=[
0,4,2,2;
2,6,2,2;
4,8,2,2
6,10,2,2]';
xrects(rects2,[-2:-1:-5])
E=gce();// compound entity
E.children.thickness=4;// thickness
```

Note that in this example the parameters `x,y,w,h` for each rectangle are listed on a line. The transpose `'` located in the definition of `rects1` and `rects2` formats the data for each rectangle as columns!

Figure 22.2 : *Drawing rectangles with* `xrects`

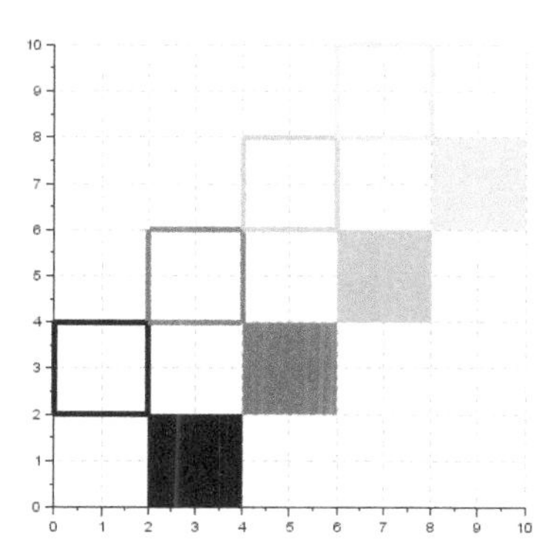

22.2. Ellipses

Defining an ellipse is similar to doing so for a rectangle. It relies on data from six real numbers `x,y,w,h,a1,a2` where:

- `x,y,w,h` define the rectangle that bounds the ellipse.
- `a1,a2` delimit the arc segment defined as follows: in an orthonormal coordinate system, with its origin at the ellipse center, the arc starts at polar angle a1*64 and ends at polar angle (a1+a2)*64, rotating counterclockwise.

Similar to rectangles, you will use the `xarc` or `xfarc` functions, which take the six parameters `x,y,w,h,a1,a2` as inputs. The `xarc` function draws only an ellipse arc, while `xfarc` sweeps an ellipse sector with a color designated by its number (refer back to the table in Figure 20.10).

```
clf;
plot2d(0,0,0,rect=[0,0,10,10],frameflag=3)
xgrid(4)
//  1st ellipse - black
xfarc(0,4,2,4,64*180,64*180)
// second ellipse (circle since inside a square)
xarc(2,4,2,2,64*0,64*360)
E=gce();
E.foreground=color("red"); //  red
// third ellipse (a circle since inside a square)
xarc(4,6,2,2,0,64*360)
E=gce();
E.foreground=color("blue"); //  blue
E.line_style=3; //  dotted line
E.thickness=5; //  thickness
// fourth ellipse
xfarc(6,8,4,2,64*45,64*270)
E=gce();
E.background=7;
```

The two functions `xarc` and `xfarc` create an Arc handle, and the `fill_mode` property controls the way in which the ellipse will get colored (filled or outlined depending on where the value is set to `"on"` or `"off"`) by using a color designated inside either the `foreground` or `background` properties.

Figure 22.3 : *Plotting ellipses with* `xarc` *and* `xfarc`

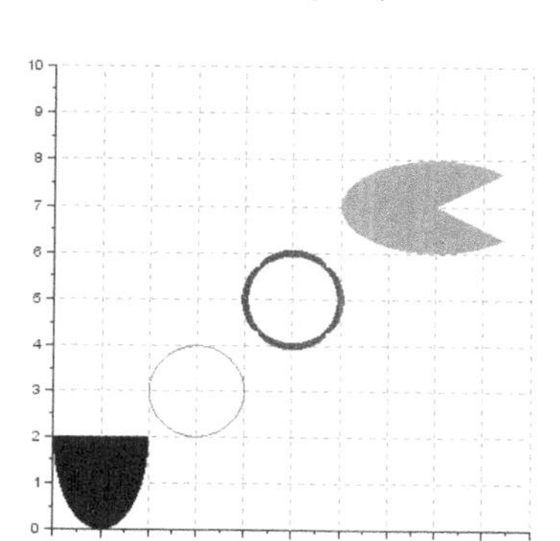

Caution › The angles that define the ellipse arc is not given in degrees or radians, but in 1/64th of a degree! This is the reason why the angle values provided are automatically multiplied by 64 in the scripts seen in this section.

In order to plot several arcs or ellipse sectors with one command, use `xarcs` or `xfarcs` which take two arguments as input:

- a matrix where each column corresponds to the parameters `x,y,w,h,a1,a2` of an ellipse
- a `fill` vector that contains the color numbers to use

You can execute the following script to get the results in Figure 22.4.

```
clf;
plot2d(0,0,0,rect=[0,0,10,10],frameflag=3)
xgrid(4)
// plotting with full ellipses
arcs1=[
0,4,2,2,0,64*360;
2,6,2,2,0,64*270;
4,8,2,2,0,64*180
6,10,2,2,0,64*90]';
xarcs(arcs1,[2:5])
// plotting rectangle outlines
arcs2=[
2,2,2,2,0,64*360;
4,4,2,2,0,64*270;
6,6,2,2,0,64*180
8,8,2,2,0,64*90]';
xfarcs(arcs2,[5:-1:2])
```

Figure 22.4 : *Drawing ellipses with* `xarcs`

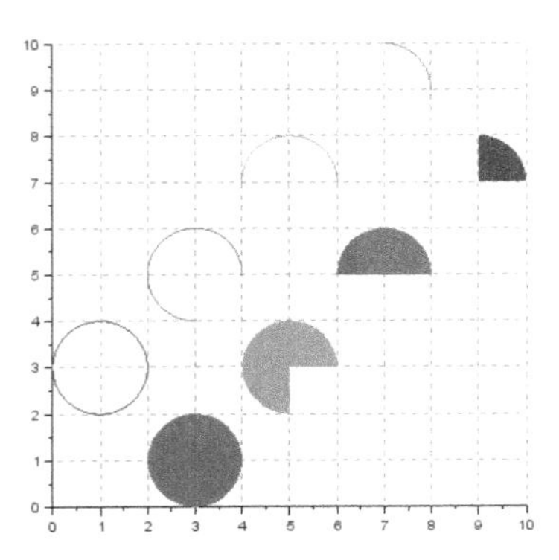

Caution › Ellipse axes drawn with Scilab are systematically oriented with respect to the direction of the coordinate system axes (horizontal and vertical).

22.3. Polygons

In general, you can draw polygons in Scilab in a way that is similar to generating rectangles and ellipses. A polygon is first and foremost a broken line that can connect to itself. We could use `plot` or `plot2d` to create it, but it is much more convenient to use the command `xpoly`. As it does with broken lines, `xpoly` takes as input two vectors `x,y` that provide the coordinates of the polygon's points, and also accepts two additional optional parameters:

- a string `"lines"` or `"marks"` to draw the segments or markers at the vertices
- an integer of value 1 for a closed polygon, or 0 otherwise

According to the option chosen, i.e. `"lines"` or `"marks"`, the `line_mode` or `mark_mode` property of the Polyline handle associated with the plot will take the value `"on"` (and the other will be `"off"`). You can then modify `line_style` or `mark_style` to change the appearance of the plot. The following script returns Figure 22.5.

```
clf;
plot2d(0,0,0,rect=[0,0,10,10],frameflag=3)
xgrid(4)
// pentagon vertices
t=[0:4]*2*%pi/5;
x=2*cos(t);
y=2*sin(t);
```

```
// 1st pentagon center at  (2.5,2.5)
xpoly(2.5+x,2.5+y)  // open polygon
E=gce();E.foreground=5;   // red
// 2nd pentagon center at (2.5,7.5)
xpoly(2.5+x,7.5+y,"lines",1)  // 1=closed polygon
E=gce();E.line_style=3;   // dotted line
// 3rd pentagon centered at (7.5,2.5)
xpoly(7.5+x,2.5+y,"marks")  // polygon vertices
E=gce();E.mark_style=2;  // dotted line
// 4th pentagon centered at (7.5,7.5)
xpoly(7.5+x,7.5+y,"lines",0)  // 0=open polygon
E=gce();E.thickness=3;  // line thickness
```

If you display the `closed` property, you can check that it is indeed this input which determines if the polygon displayed is closed or not according to its value (0 ou 1).

Figure 22.5 : *Drawing Polygons with* `xpoly`

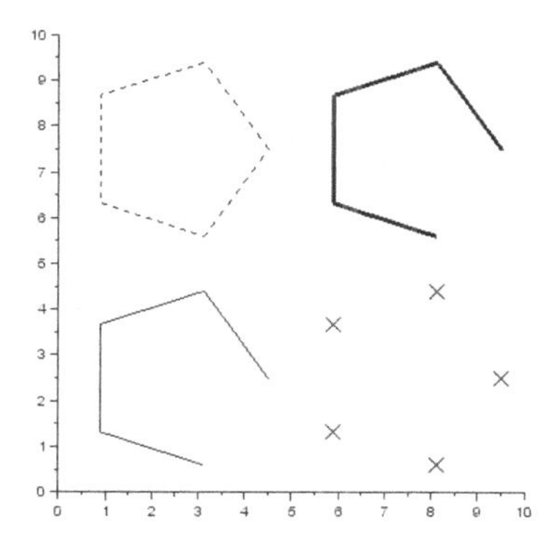

Drawing a filled polygon

To draw a polygon filled with a color, use the command `xfpoly`. In addition to the two lists of coordinates for the polygon's points, the command takes as input a numerical argument `fill` that satifies the following:

- If `fill>0` the polygon is filled with the `fill` color and its outline is drawn with the current color and line style.
- If `fill<0` the polygon is filled with the `-fill` color and its contour is not drawn.
- If `fill=0` or if it is omitted, the polygon's interior gets filled with the current color along with its outline.

Execute the following script to get Figure 22.6 .

```
clf;
plot2d(0,0,0,rect=[0,0,10,10],frameflag=3)
xgrid(4),
// pentagon vertices
t=[0:4]*2*%pi/5;
x=2*cos(t);
y=2*sin(t);
// 1st pentagon centered at (2.5,2.5)
xfpoly(2.5+x,2.5+y)  // black background
E=gce();E.foreground=5;   // red edge
// 2nd pentagon centered at (2.5,7.5)
xfpoly(2.5+x,7.5+y,5)  // red background
E=gce();E.line_style=3;   // edge is a black dotted line
// 3rd pentagon centered at (7.5,2.5)
xfpoly(7.5+x,2.5+y,-2)  // polygon vertices
E=gce();E.mark_style=2;  // dotted line
// 4th pentagon centered at (7.5,7.5)
xfpoly(7.5+x,7.5+y,0)  // 0=open polygon
E=gce();E.background=3;  // green background
```

If you look at the polygon's Polyline handle, you will see that the `fill_mode` property is set to `"on"` and the color used is stored in the `background` property (for the inside) and `foreground` property for the contour.

Figure 22.6 : *Drawing polygons with* `xfpoly`

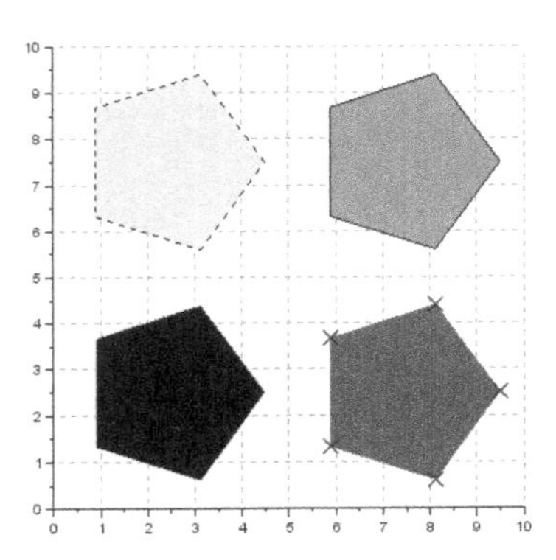

Tip › Starting from polygons, it is possible to create any geometrical figure as long as you account for enough sides.

The following script yields the figure **below**.

```
clf;
plot2d(0,0,0,rect=[-1.5,-1.5,1.5,1.5],frameflag=3)
xgrid(4),
```

```
// drawing a contour
t=[0:0.02:2*%pi]';
r=1+0.2*sin(10*t);
x=r.*cos(t);
y=r.*sin(t);
xfpoly(x,y,5)
A=gca();A.isoview="on";  // isometric scale
```

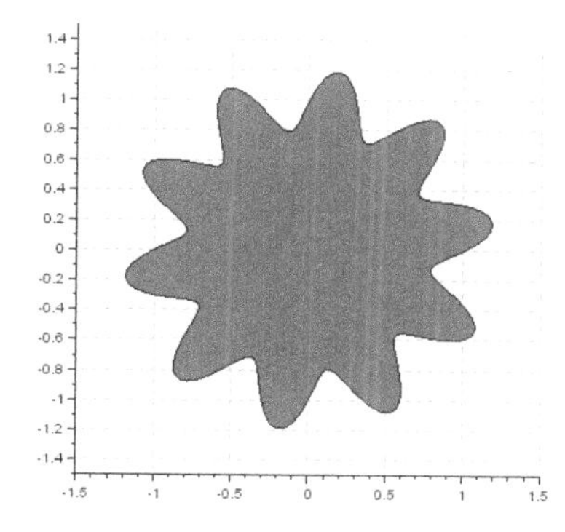

Drawing several polygons

To plot several polygons, you will need to use the functions `xpolys` or `xfpolys`. These command take as arguments:

- two `pxn` matrices whose columns provide the coordinates `x` and `y` of the vertices of a polygon with `p` vertices
- a list of `n` numbers for the color of each polygon

Execute the following script to get Figure 22.7.

```
clf;
plot2d(0,0,0,rect=[0,0,10,10],frameflag=3)
xgrid(4),
// pentagon vertices
t=[0:4]*2*%pi/5;
x=2*cos(t)';
y=2*sin(t)';
X=[2.5+x 2.5+x 7.5+x 7.5+x];
Y=[2.5+y 7.5+y 2.5+y 7.5+y];
// edges of the four black pentagons
xpolys(X,Y,[1 1 1 1])
E=gce();E.children.closed=1;  // closed boundary
E.children.thickness=3;       // thickness
// 4 pentagons filled with color
```

```
xfpolys(X,Y,[2 3 5 7])
```

Note that in this script each polygon corresponds to one column of `X` and one of `Y`. The different polygons created are gathered together inside a Compound handle and retrieve in the variable `E`. `E.children.closed` returns the properties of `E`'s four children.

Figure 22.7 : *Plotting several polygons with `xpolys` and `xfpolys`*

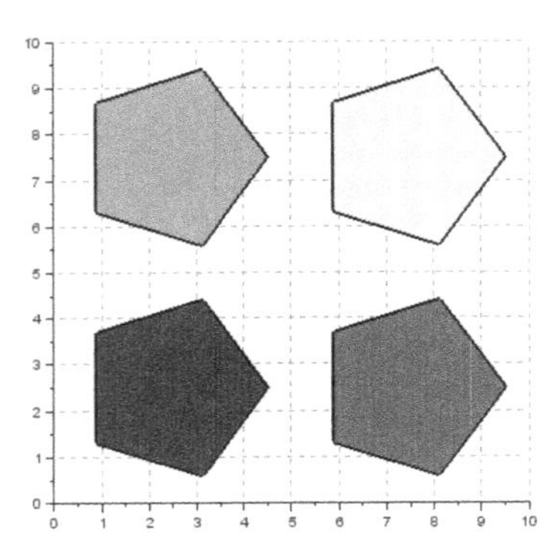

*Caution › By default, the polygons created with **`xpolys`** are not closed. In order for the contours to get displayed, you need to change the handles' **`closed`** value to **`1`**.*

Drawing a regular polygon

The command `xrpoly` lets you create a regular polygon, by providing:

- the coordinates `x, y` of its center
- the number of sides
- its diameter (that of its circumscribed circle)

One last (optional) parameter lets you rotate the polygon around its center by a given angle (in radians) in the positive direction. Execute the following script to generate Figure 22.8.

```
clf;
plot2d(0,0,0,rect=[0,0,10,10],frameflag=3)
xgrid(4),
// regular hexagon
xrpoly([5,5],6,4)
E=gce();
E.foreground=5;  // color of the red contour
E.thickness=3;    // contour thickness
// 2nd smaller hexagon rotated 90 degrees
xrpoly([5,5],6,2,%pi/2)
```

Figure 22.8 : *Drawing regular polygons with* `xrpoly`

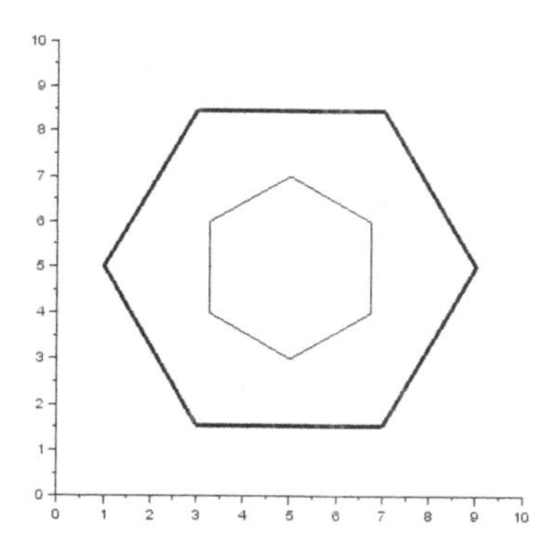

22.4. Arrows and segments

You can also draw disconnected segments, unlike those of the Polyline handle. To do this, use the command `xsegs` which takes as input two `2xn` matrices with the coordinates of the starting point (first row) and end point (second row) for each of the `n` segments. A third argument lets you associate a color to each segment. Execute the following script to get Figure 22.9.

```
clf;
plot2d(0,0,0,rect=[0,0,10,10],frameflag=3)
xgrid(4)
// segments start
x1=2*ones(1,5)
y1=[1:2:9];
// segments end
x2=8*ones(1,5)
y2=[1 1 5 9 7];
// drawing the segments
X=[x1;x2],Y=[y1;y2]
```

```
xsegs(X,Y,[1:5])
E=gce();
E.thickness=3;
E.mark_size=2;
E.mark_mode="on";
```

Segments belong to an entity of type *segs* which has properties of type `mark_*` that let you deal with the markers' appearance and segment's endpoints.

Figure 22.9 : *Drawing segments with* `xsegs`

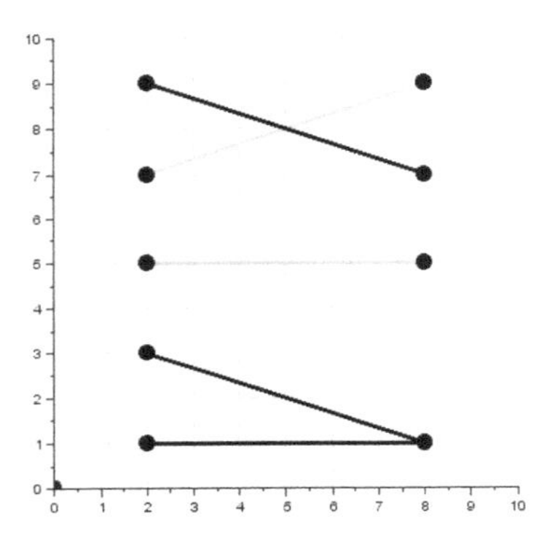

You can also just use the function `xarrows` to generate arrows. This command is similar to `xsegs` but it also take as argument an integer indicating the size of the arrows. Execute the following script to get Figure 22.10.

```
clf;
plot2d(0,0,0,rect=[0 0 10 10],frameflag=3)
xgrid(4),
// segments start
x1=2*ones(1,5)
y1=[1:2:9];
// segments end
x2=8*ones(1,5)
y2=[1 1 5 9 7];
// drawing the segments
X=[x1;x2],Y=[y1;y2]
xarrows(X,Y,3,[1:5])
E=gce();
E.thickness=3;
```

The third argument of `xarrows` sets the arrow size to 3. In fact, the arrows' `arrow_size` property controls the size of the arrows drawn. This handle is of the type segs so the arrows are just segments for which the value of `arrow_size` is different from 0.

Figure 22.10 : Drawing arrows with `xarrows`

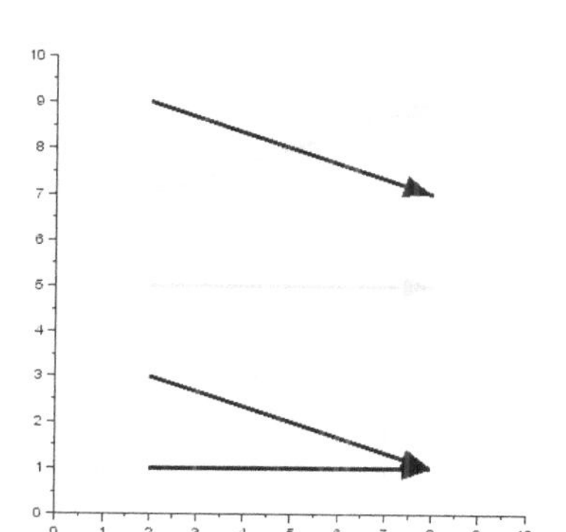

22.5. Vector fields

A special case of arrows is vector fields. A vector field is a function which associates a vector (of coordinates `(u,v)`) to each point (of coordinates `(x,y)`) in the plane. Portraying a vector field graphically involves drawing an arrow symbolizing the vector of coordinates `(u,v)` by locating it at the associated point `(x,y)`. A field is consequently defined by four matrices:

- two column matrices `x` and `y` (sizes `p` and `n`) which discretize the region covered by the field
- two matrices `fx` and `fy` of size `pxn` such that the vector associated to the point of coordinates `(x(i),y(j))` has the coordinates `u=fx(i,j)` and `v=fy(i,j)`

In Scilab, vector fields are figures defined by handles belonging to the *champ* type. The `champ` function lets you automatically plot a vector field from four matrices `x,y,fx,fy`. This `champ` function additionally admits several optional arguments of the same type as those admitted by `plot2d` (see Section 20.3, *plot2d command and other types of plots*). Execute the following script to get Figure 22.11.

```
clf; xgrid(1)  // grid
A=gca();A.isoview="on";  // isometric display
x=[0:10]';y=x;  // discretizing x,y data
fx=grand(11,11,'def');fy=grand(11,11,'def');
champ(x,y,fx,fy,rect=[0,0,10,10])
```

The arrows displayed are in the same direction as the vector defined by `(fx,fy)` and the length is proportional. These data are stored inside the following Champ handle properties : data.x, data.y, data.fx, data.fy. Here we have added a fifth argument to champ to set the size of the graphics window.

Figure 22.11 : *Plotting vector fields with* champ

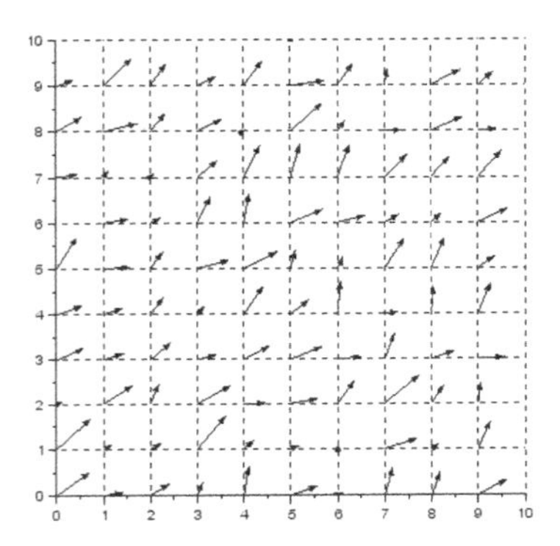

The champ1 function lets you display arrows of the same length at each point (x,y) but whose colors vary directly proportionally to the length of the vector defined by (fx,fy). This color is taken from the color table associated to the Figure handle in which the plot is created (see Figure 21.10). Execute the following script to get Figure 22.12.

```
clf; xgrid(1)  // grid
A=gca();A.isoview="on";  // isometric display
x=[0:10]';y=x;  // discretizing x,y data
fx=grand(11,11,'def');fy=grand(11,11,'def');
champ1(x,y,fx,fy,rect=[0,0,10,10])
// choose a different color table
F=gcf();F.color_map=jetcolormap(64)
```

You can check that the handle of type *champ* has a property called colored which here is set to "on". This means the arrows get displayed in color and with the same length (i.e. without using data.fx,data.fy). Furthermore, we have chosen a table of colors with 64 shades with jetcolormap.

Figure 22.12 : *Plotting vectors fields with* `champ1`

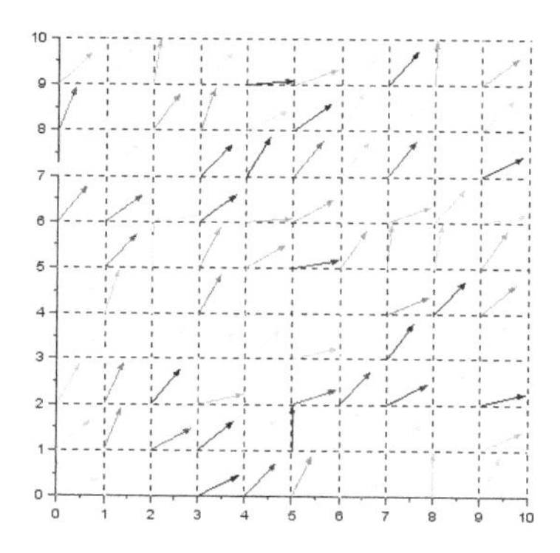

In practice, vector fields are used to display systems of differential equations of the form:

$$\frac{d}{dt}\begin{pmatrix} x \\ y \end{pmatrix} = f(t, x, y) \qquad \text{with} \qquad \begin{aligned} f:\ & R \times R^2 && \longrightarrow R^2 \\ & (t, (x, y)) && \longmapsto (u, v) \end{aligned}$$

It is thus more convenient to display a vector field by directly using the function *f*. You can achieve this with the function `fchamp` which takes as input:

- the function `f`
- the value of `t`
- the two vectors `x` and `y` (same as for the functions `champ` and `champ1`)

Execute the following script to generate Figure 22.13:

```
clf; xgrid(4)  // grid
A=gca();A.isoview="on";  // isometric display
// choose a different color table
F=gcf();F.color_map=jetcolormap(64)
// two functions for vector fields
function [u]=converge(t,x)
    u(1)=-x(1)
    u(2)=-x(2)
endfunction
function [u]=rotation(t,x)
    u(1)=-x(2)
    u(2)=x(1)
endfunction
```

```
// plot the vector fields
x=[-4:4]';y=x;
rect=[-4 -4 4 4]
// vector field with current color table
fchamp(converge,0,x,y,rect=rect)
E=gce();E.colored="on";
// vector field in black
fchamp(rotation,0,x,y,rect=rect)
```

In this example, we have generated two different vector fields:

- one associated to the `rotation` function
- another associated to the `converge` function

with the second one colored with the current color table by changing the `colored` property of the corresponding Champ handle.

Figure 22.13 : *Plotting vector fields with* `fchamp`

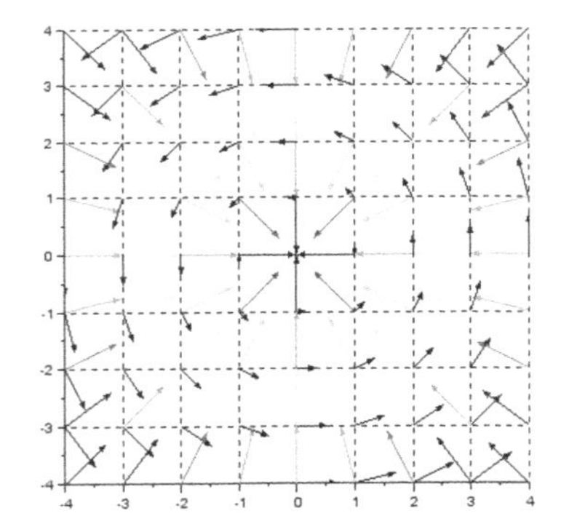

22.6. Histograms and other charts

In statistics, we encounter numerous types of fairly complex charts and histograms. Scilab has several functions that can be used to automatically generate these charts from raw data. Graphics entities that make up these charts are in general composed of several handles gathered inside a Compound handle. In the examples shown in this section, you will use the `grand` command to generate a collection of (pseudo-)random data (see Part *Computing*) as shown in the following script.

```
-->// generate a list of random values

-->X=grand(10,1,'bin',10,0.4)
 X  =
    1.
    3.
    5.
    3.
    4.
    2.
    4.
    5.
    2.
    1.

-->m=tabul(X)    //table of frequencies
 m  =
    5.    2.
    4.    2.
    3.    2.
    2.    2.
    1.    2.

-->x=m(:,1)      //values
 x  =
    5.
    4.
    3.
    2.
    1.

-->n=m(:,2)      //counts
 n  =
    2.
    2.
    2.
    2.
    2.
```

The `tabul` function lets you extract values and associated counts for a whole statistical set. You can also use the function `dsearch` to analyze lists of data that need to be binned in intervals.

Generating bar charts

To start off, you can create bar charts by using the commands `bar` and `barh`, which takes as input the values `x` and the counts `n` corresponding to a statistical data set.

```
// generating a list of random values
X=grand(100,1,'bin',10,0.4);
m=tabul(X)    // table of frequencies
```

```
x=m(:,1)      // values
n=m(:,2)      // bin counts
clf()
subplot(121)  // vertical bars
bar(x,n)
subplot(122)  // horizontal bars and options
barh(x,n,0.5,'g')
```

The previous scripts yields Figure 22.14. The difference between the two functions is that `barh` plots horizontal bars whereas `bar` plots vertical ones. The two functions take the same optional arguments:

- a number between 0 and 1 to adjust the length of bars (a percentage of the maximum length)
- the first letter of the color name to define the color of the bars (same as for the `plot` function see Figure 20.2)

Figure 22.14 : *Plotting bar charts with* `bar` *and* `barh`

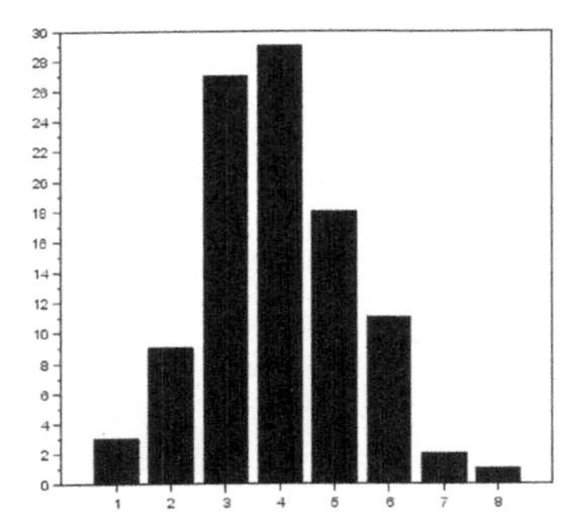

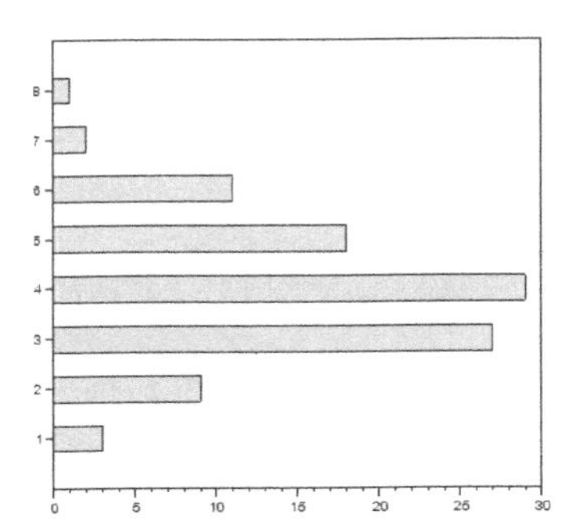

Creating grouped or stacked bar charts

You can also create more complex charts that show several bars for each value by adding the option:

- `'grouped'` (or leave empty) for side-by-side bars
- `'stacked'` for bars on top of each other

Execute the following script to get Figure 22.15.

```
// generate a list of random values
x=[1:10]';
n1=x;n2=11-x;n3=5*ones(x);
clf()
subplot(121)  // vertical grouped bars
bar(x,[n1 n2 n3],'grouped')
subplot(122)  // horizontal stacked bars
barh(x,[n1 n2 n3],0.5,'stacked')
```

You may notice that we only need to provide the vector of values (`x`) once and the lists of counts are the columns in the second matrix (`[n1,n2,n3]`). The colors used for the different bars are given in the color table of the associated Figure handle.

Figure 22.15 : *Plotting grouped or stacked bar charts with* `bar` *and* `barh`

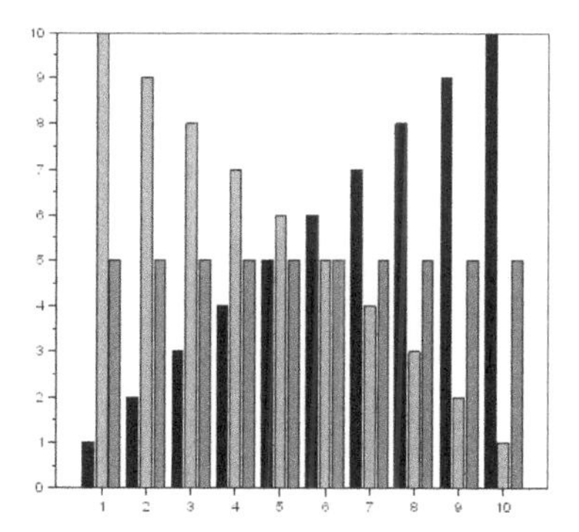

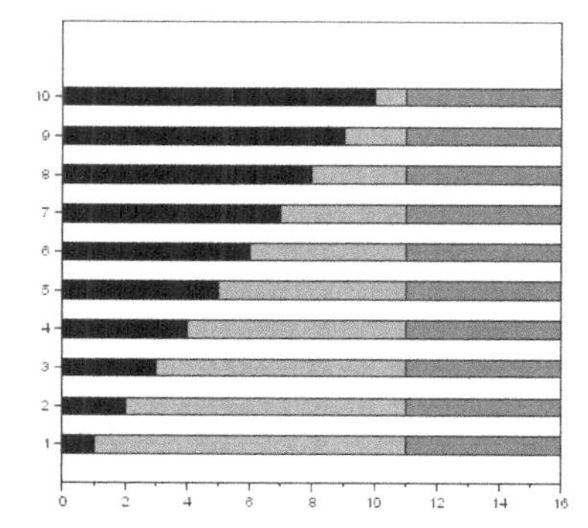

Creating pie charts

If you'd rather create pie charts, use the command `pie`. This function takes the bin counts as arguments to draw the pie. We can then provide additional optional arguments to change its appearance:

- a vector with 0 or 1 to indicate which parts of the pie to emphasize
- a strings vector with the text that needs to be displayed next to each slice (by default the percentage associated to each slice)

Execute this script to get Figure 22.16.

```
// generate a list of random values
X=grand(100,1,'bin',4,0.4);
m=tabul(X)     // table of frequencies
x=m(:,1)       // values
n=m(:,2)       // bin counts
clf()
subplot(121)  // simple pie chart
pie(n)
subplot(122)  // exploded pie chart
pie(n,bool2s(n==max(n)),'x='+string(x))
```

For the second pie chart, we have displayed the value of the count (by translating it to a string with the command `string`) and we have emphasized the slice with the largest count with the help of `bool2s(n==max(n))` (see the `bool2s` command in chapter *Booleans*).

Figure 22.16 : *Plotting pie charts with* `pie`

Creating histograms covering given intervals

If you need to work with a set of numbers that you wish to group into intervals, use the command `histplot`. This command takes two matrices as inputs:

- The first lets you define the bins (the intervals used to arrange the data):
  - either with a vector containing the interval's limits
  - or with an integer providing the number of intervals
- The second contains the data set.

Execute the following script to generate Figure 22.17.

```
// generate a list of random values
X=grand(100,1,'def');
clf()
subplot(121)  // histogram with 5 bins
histplot(5,X)
subplot(122)  // histogram with variable size bins
histplot([0 0.2 0.5 0.6 0.7 1],X)
```

Figure 22.17 : *Plotting histograms with* `histplot`

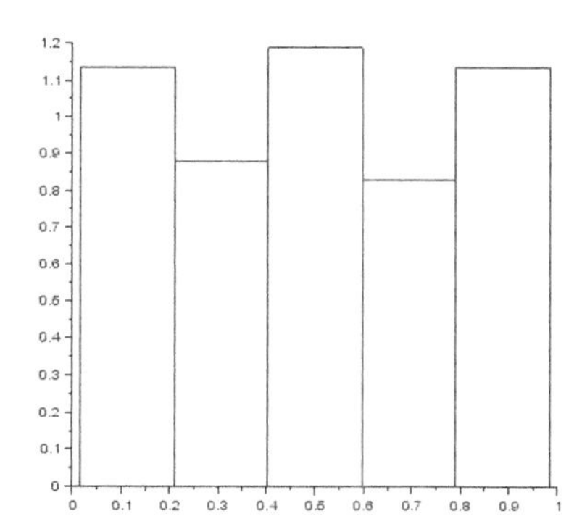

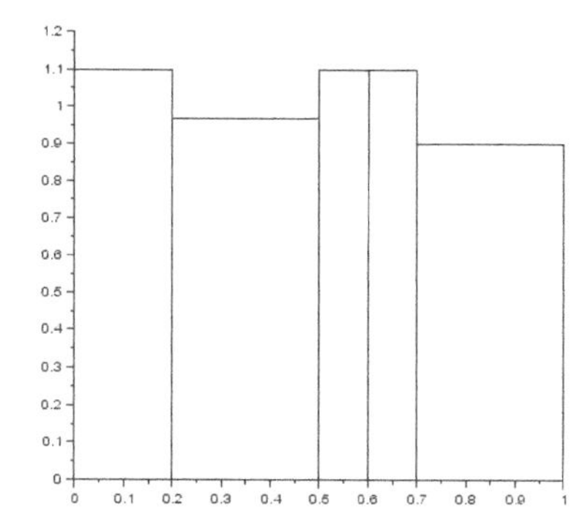

The `histplot` command can also take an optional argument as a boolean value:

- `normalization=%f`: the height of each rectangle is proportional to the count associated with the interval.
- `normalization=%t`: the surface of each rectangle is proportional to the count associated with the interval (default setting).

The y-axis graduations are set as a function of this normalization. You can check this with the following script which creates Figure 22.18.

```
// generate a list of random values
X=grand(100,1,'nor',5,2);
clf()
subplot(121)  // normalized histogram
histplot([0 2 4 5 7 10],X,normalization=%t)
subplot(122)  // non-normalized histogram
histplot([0 2 4 5 7 10],X,normalization=%f)
```

Figure 22.18 : Normalized or non-normalized histogram with `testhistplot`

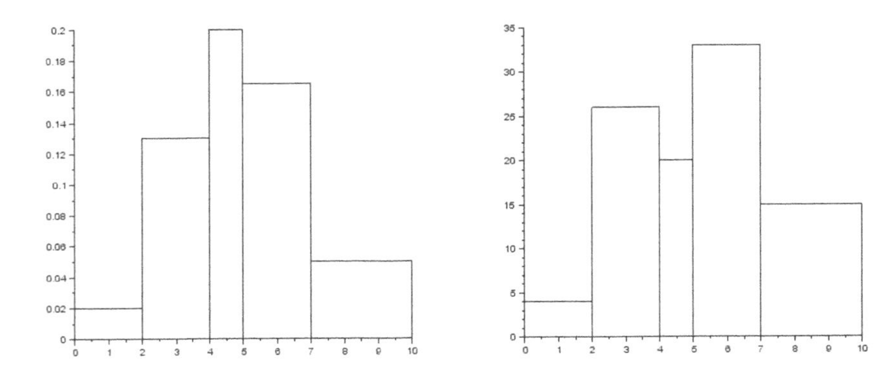

Creating 3D histograms

Finally, you can create 3D histograms by using `hist3d`. In the following script, we provide a matrix as argument in which the data `Z(i,j)` determine the height of the column located at `(i-1,j-1)`.

```
Z=([5:-1:1]')*ones(1,5)+ones(5,1)*[5:-1:1]
clf(); hist3d(Z)
```

Figure 22.19, generated by this script, is made from a Fac3d handle. The `hist3d` function takes as input optional arguments of the same type as `plot3d` (see Section 21.4, *Plotting functions of two variables*).

Figure 22.19 : Three-dimendional histogram with `hist3d`

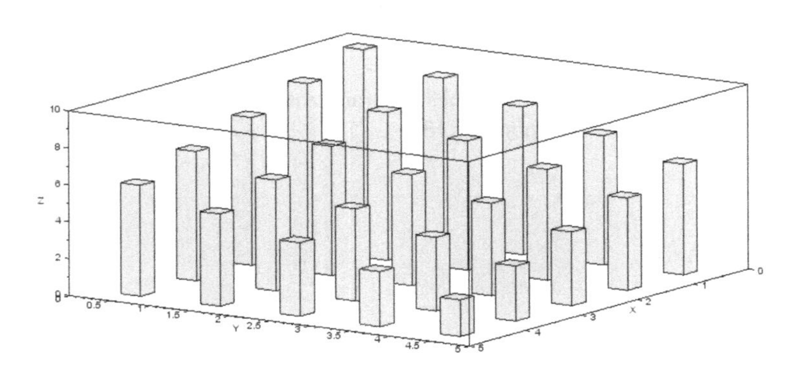

23

To Go Even Further

Scilab's capabilities regarding graphics is not limited to plots of broken lines and facets. You also have the option of adding text to a figure, of animating it and even making it interactive by adding buttons, fields to fill out or making it sensitive to mouse clicks.

23.1. Adding text to figures

The `xstring` command lets you add text to a figure in various forms and in different positions in the figure. This function creates Text handles which are often added to other entities inside Compound handles. To display a string, in addition to the `str` string, you need to at least provide the position `x,y` of the starting point for the string display by following the format `xstring(x,y,str)`. You can also provide two optional arguments:

- an angle to rotate the string (clockwise)
- an integer (1 or 0) to draw a box around the string of characters

For example, the following script displays the character string in Figure 23.1.

```
// launch the graphics window
clf;
plot2d(0,0,0,rect=[0,0,1,1])
xgrid(4)
// the string to display
alpha="abcdefghijklmnopqrstuvwxyz"
// Courier font, black
xstring(0.1,0.25,alpha)
E=gce()
E.font_size=1
E.font_style=0
E.font_foreground=color("black")
// Symbol font, blue, boxed
xstring(0,0,alpha,0,1)
E=gce()
E.font_size=4
E.font_style=1
E.font_foreground=color("blue")
// Times font, black, slanted
xstring(0.2,0.5,alpha,20)
E=gce()
```

```
E.font_size=5
E.font_angle=20
E.font_style=2
E.foreground=2
// Times font, Bold over red background
xstring(0.3,0.7,alpha,340)
E=gce()
E.font_size=3
E.font_style=4
E.box="on"
E.fill_mode="on"
E.background=5
```

To fine-tune the text's display, you need to modify the Text handle's properties that can be retrieved with `gce()`:

- `text` contains the character string to display.
- `position` contains the coordinates of the lower-left corner of the rectangular box around the displayed text.
- `font_foreground` lets you pick the font color for the characters *via* a numerical value and the current color table.
- `fill_mode` and `background` let you change the font color.
- `font_size` lets you choose the character size by specifying a value between 0 and 10 which will produce characters of the sizes 8 pt, 10 pt, 12 pt, 14 pt, 18 pt, 24 pt, 30 pt, 36 pt, 42 pt, 48 pt or 54 pt, respectively.
- `font_style` lets you choose a given font by specifying a value between 0 and 9 which selects one of the following fonts Courier, Symbol, Times, Times Italic, Times Bold, Times Bold italic, Helvetica, Helvetica Italic, Helvetica Bold, Helvetica Bold italic (also see the script and figure in the following section).
- `text` lets you box (or not) a character string.

Figure 23.1 : Displaying text in a graphics window with `xstring`

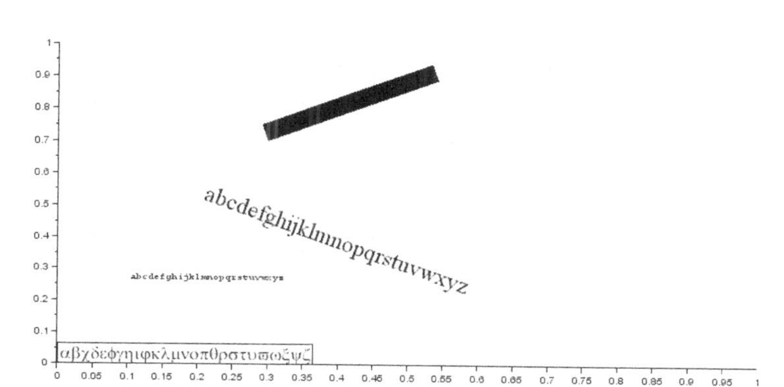

Choosing a font

The following script lists the 10 different available fonts (see Figure 23.2).

```
clf();
// change font
xstring(0.15,1,'Fonts using font_style')
xstring(0,0.8,'0 Courier')
E=gce();E.font_style=0;
xstring(0.5,0.8,'1 Symbol')
E=gce();E.font_style=1;
xstring(0,0.6,'2 Times')
E=gce();E.font_style=2;
xstring(0,0.4,'3 Times italic')
E=gce();E.font_style=3;
xstring(0,0.2,'4 Times Bold')
E=gce();E.font_style=4;
xstring(0,0,'5 Times Bold italic')
E=gce();E.font_style=5;
xstring(0.5,0.6,'6 Helvetica')
E=gce();E.font_style=6;
xstring(0.5,0.4,'7 Helvetica')
E=gce();E.font_style=7;
xstring(0.5,0.2,'8 Helvetica Bold')
E=gce();E.font_style=8;
xstring(0.5,0,'9 Helvetica Bold italic')
E=gce();E.font_style=9;
// change font size
A=gca();A.children.font_size=5;
```

Figure 23.2 : *Available fonts via* `font_style`

Fonts using font_style

0 Courier	1 Σψμβολ
2 Times	6 Helvetica
3 Times italic	*7 Helvetica*
4 Times Bold	**8 Helvetica Bold**
5 Times Bold italic	***9 Helvetica Bold italic***

If you find this list of fonts too limited, you can load other fonts stored in your system by calling `xlfont`. For example, the following script yields Figure 23.3.

```
// list of available fonts
fonts=xlfont('AVAILABLE_FONTS');
// load the font French Script MT
xlfont('French Script MT',10)
// display text with French Script MT
clf;
xstring(0.1,0.75,'Times font in Scilab 5.5.0')
E=gce();E.font_style=2; E.font_size=5;
xstring(0.1,0.5,'Times Bold Italic font in Scilab 5.5.0')
E=gce();E.font_style=5; E.font_size=5;
xstring(0.1,0.25,'French Script MT font in Scilab 5.5.0')
E=gce();E.font_style=10; E.font_size=6;
```

Figure **23.3 :** *Loading new fonts with* `xlfont`

Times font in Scilab 5.5.0

Times Bold Italic font in Scilab 5.5.0

French Script MT font in Scilab 5.5.0

Tip › *Depending on the fonts installed in your system, you have several possible choices of* **`font_size`** *which can exceed the limit of 10* ***default fonts****. These fonts are most often stored inside files of extension* `*.ttf` *or* `*.otf` *inside the directory:*

- `C:\WINDOWS\Fonts` *for Windows*
- `~/.fonts` *for Linux*
- `/System/Library/Fonts` *for Mac OS*

Inserting LaTeX or MathML formulas

In addition to the fonts already mentioned, you have the option of writing formulas in LaTeX, or MathML, and pass it directly as argument for the `xstring` command, as shown in the following example.

```
clf();
// insert LaTeX formulas
xstring(0.5,1,'$\LaTeX $')
xstring(0.75,0.75,'${\pi\over 2}=\int_{0}^\infty {\sin(x)\over x}\; dx
$')
xstring(0.25,0.75,'$${\pi^2\over 6}=\sum_{n=1}^\infty {1\over n^2}$$')
xstring(0.25,0.25,prettyprint(eye(4,4)))
xstring(0.75,0.25,'$g(x)=$ '+prettyprint(poly(1,'x')/(2*poly(0,'x')-1)))
xstring(0.25,0,'$f(x)=$ '+prettyprint(poly(1:4,'x')))
// change font size
A=gca();A.children.font_size=5;
```

If you are familiar with the LaTeX language, you will easily recognize the code corresponding to the formulas displayed in Figure 23.4. You will note that we have used the command `prettyprint` (see Chapter *Character Strings and Text Files*) to immediately convert certain Scilab variables (here a matrix and polynomial) into LaTeX code.

Figure 23.4 : *Displaying mathematical formulas with LaTeX*

LaTeX

$$\frac{\pi^2}{6} = \sum_{n=1}^{\infty} \frac{1}{n^2} \qquad \frac{\pi}{2} = \int_0^{\infty} \frac{\sin(x)}{x}\, dx$$

$$\begin{pmatrix} 1 & 0 & 0 & 0 \\ 0 & 1 & 0 & 0 \\ 0 & 0 & 1 & 0 \\ 0 & 0 & 0 & 1 \end{pmatrix} \qquad g(x) = \frac{-1 + x}{-1 + 2x}$$

$$f(x) = 24 - 50x + 35x^2 - 10x^3 + x^4$$

Tip › *When you enter strings between the characters* **`$...$`** *inside the* **`scinotes`** *editor, the LaTeX rendering of the formula gets displaying it inside a tooltip as shown in the figure* **below** *when the mouse hovers over it.*

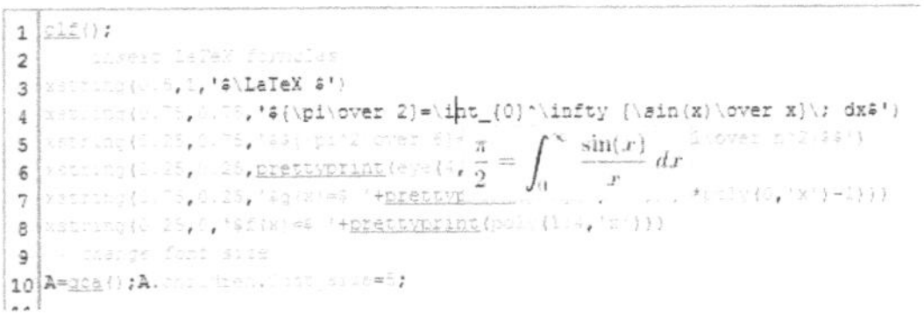

Caution › *Mixing text and LaTeX formulas inside one string of characters displayed with* **`xstring`** *or inside a strings matrix displayed with* **`titlepage`** *can be challenging. For example:*

```
clf();
// inserting LaTeX formulas
xstring(0,0.9,'isolated text  and only one formula ')
xstring(0.6,0.8,'${\pi^2\over 6}=\sum_{n=1}^\infty {1\over n^2}$')
xstring(0,0.5,'text and formulas ${\pi^2\over 6}=\sum_{n=1}^\infty
 {1\over n^2}$')
xstring(0,0,'$\textrm{text and formulas together but inside \$\dots \
$} : {\pi^2\over 6}=\sum_{n=1}^\infty {1\over n^2}$')
// change font size
A=gca();A.children.font_size=5;
```

isolated text and only one formula $\frac{\pi^2}{6} = \sum_{n=1}^{\infty} \frac{1}{n^2}$

text and formulas ${\pi^2\over 6}=\sum_{n=1}^\infty {1\over n^2}$

text and formulas together but inside $\$ \ldots \ \$: \frac{\pi^2}{6} = \sum_{n=1}^{\infty} \frac{1}{n^2}$

Positioning text

When you want to correctly locate strings in a figure, the difficulty lies in evaluating how much space the string will take. The space occupied can be viewed as the smallest rectangular box that bounds the string of characters. We can define this rectangle as follows:

- the coordinates `x,y` of the corner of a rectangle (the lower-left one for example) along with its width `w` and height `h`
- or the coordinates of the four corners of the rectangle that bounds the string

You can retrieve this data by using the commands `xstringl` (for `[x,y,w,h]`) or `stringbox` (for the coordinates of the four corners) as shown in the following example Figure 23.5.

```
-->// launch the graphics window

-->clf;

-->plot2d(0,0,0,rect=[0,0,1,1])

-->xgrid(4)

-->// first example

-->txt="the formula below is lopsided!"
 txt  =
 the formula below is lopsided!
```

```
-->xstring(0.1,0.75,txt)

-->E=gce();

-->E.font_size=5;

-->E.font_style=3;

-->E.box="on";

-->rect=xstringl(0.1,0.75,txt,3,5)
 rect  =
    0.1
    0.8514493
    0.6557377
    0.1014493

-->corners = stringbox(txt,0.1,0.75)
 corners  =
    0.1     0.1          0.4191257    0.4191257
    0.75    0.8021739    0.8021739    0.75

-->//  second example

-->formula='${\pi^2\over 6}=\sum_{n=1}^\infty {1\over n^2}$';

-->xstring(0.4,0.25,formula)

-->E=gce();

-->E.font_angle=20;

-->E.font_size=5;

-->E.box="on";

-->rect=xstringl(0.4,0.25,formula)
 rect  =
    0.4
    0.3543478
    0.1333333
    0.1043478

-->corners = stringbox(E)
 corners  =
    0.4     0.4627971    0.7524073    0.6896102
    0.25    0.4787947    0.3390126    0.1102179
```

Figure 23.5 : *Rectangles bounding character strings with* `xstringl` *or* `stringbox`

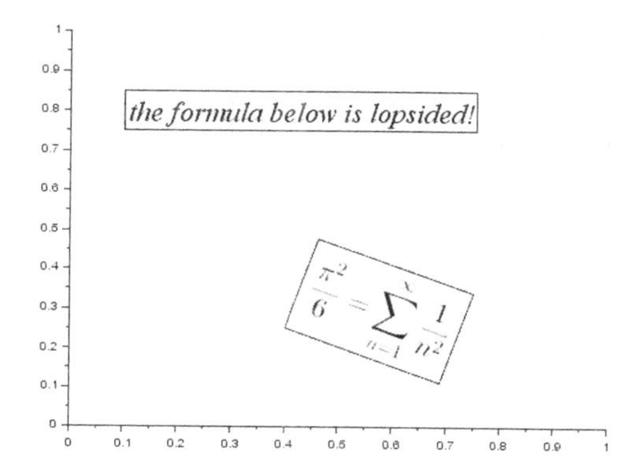

To insert text in a figure and have it centered and enclosed in a box defined by four numbers `[x,y,w,h]`, you can also use the function `xstringb`. Execute the following script:

```
// launch the graphics window
clf;
plot2d(0,0,0,rect=[0,0,1,1])
xgrid(4)
// string to display
alpha="abcdefghijklmnopqrstuvwxyz";
// size of rectangle bounding the string
rect=xstringl(0,0,alpha,3,5);
// display the rectangle
x=0.1,y=0.6,w=rect(3),h=rect(4),
xrect(x,y+4*h,1.5*w,4*h)
// display the string
xstringb(x,y,alpha,w*1.5,h*4)
E=gce();
E.font_size=5;
E.font_style=3;
E.box="on";
// size of the rectangle inside which the string is centered
rect=xstringl(0,0,alpha,3,3);
// display the rectangle
x=0.25,y=0.25,w=rect(3),h=rect(4),
X=x+[0 w w 0]';Y=y+[0 0 h h]';
xrect(x,y+3*h,1.2*w,3*h)
// display the string
xstringb(x,y,alpha,w*1.2,h*3)
E=gce();
E.font_size=3;
```

```
E.font_style=3;
E.box="on";
```

then compare the values of `x,y,w,h` to the coordinates of the corners of the boxes displayed in Figure 23.6.

Figure 23.6 : *Displaying text centered inside a box with* `xstringb`

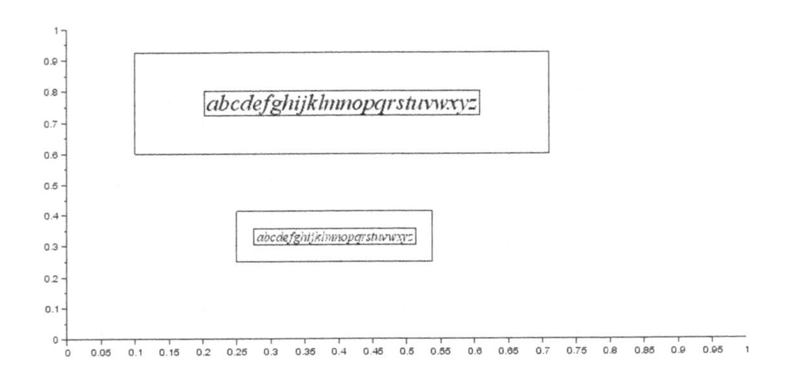

Caution › The `xstringb` *command does not display a box! You can use* `xrect` *to display it, but in this case the values of the parameters* `x,y` *that define the box bounding the string correspond to:*

- *the lower-left point for* `xstringb` *and* `xstring`
- *the upper-left for* `xrect`

Adding a title or title page

We have already seen that titles can be added to figures by using `xtitle` or `title`. Unlike `xstring`, these functions create Label handles, connected to the current Axes handle's `title` property. Execute the following script to get the title in Figure 23.7.

```
// plotting a figure
x=[1:0.01:6];y=sin(x);
clf; plot(x,y,'-r')
// add a title
title('y=sin(x)','fontsize',5,'color','red','backgroundcolor','cyan')
// Plot axes
A=gca();
// change the font
A.title.font_style=3;
```

***Figure 23.7 :** Figure title created with `title`*

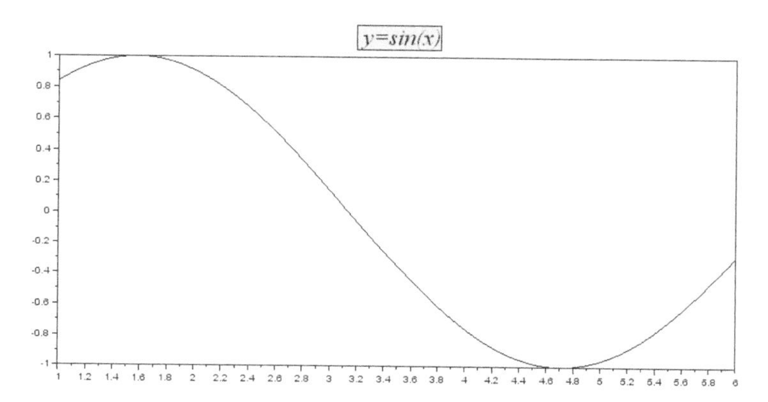

Next, look at the `title` property of the Axes handle (the Label handle) as shown here:

```
-->A.title  // Label handle properties
 ans  =
Handle of type "Label" with properties:
=======================================
visible = "on"
text = "y=sin(x)"
font_foreground = 5
foreground = -1
background = 4
fill_mode = "on"
font_style = 3
font_size = 5
fractional_font = "off"
font_angle = 0
auto_position = "on"
position = [3.0136612,1.0434783]
auto_rotation = "on"
tag =
```

You will recognize in this example most of the properties associated with text handles for Scilab figures. Similarly, you can create a title page with `titlepage` such as in Figure 23.8.

```
txt=['titlepage';
'this is a title page';
'it can span several lines';]
clf;titlepage(txt)
```

The `txt` strings matrix is located inside a Text handle, and gets displayed and centered on the page with a border. If you adjust the `alignment` property to `"left"` or `"right"`, you will get a different alignment.

Figure 23.8 : *Title page created with* `titlepage`

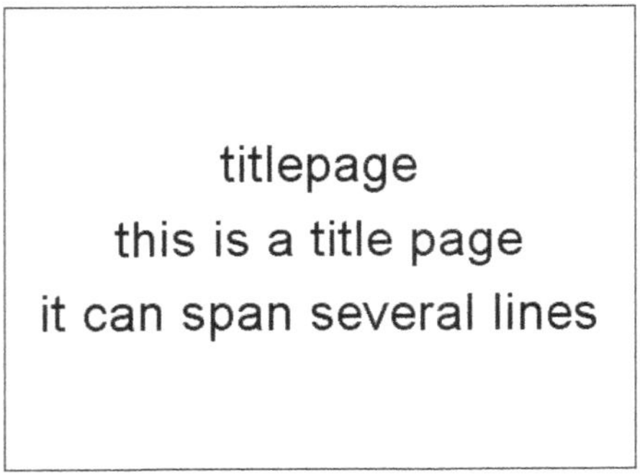

titlepage
this is a title page
it can span several lines

Caution › *Handles of type* label *(or* axis*) which are often found inside the properties of an Axes handle, are not listed inside the* **`children`** *property, even though they are descendants of the Axes handle!*

Labeling coordinate axes

To assign more descriptive labels to your coordinate axes, you need to proceed as follows:

```
// generate a list of random values
X=grand(100,1,'bin',10,0.6);
m=tabul(X);     //  table of frequencies
x=m(:,1);       //  values
n=m(:,2);       //  bin counts
clf();bar(x,n) //  histogram
A=gca();
A.x_label.text="grade over 10";
A.x_label.font_size=3;
A.x_label.font_style=4;
A.y_label.text="counts";
A.y_label.font_size=3;
A.y_label.font_style=4;
```

After executing, the histogram generated will have display the label `grade over 10` on the x-axis (see Figure 23.9). Note that the `x_label`, `y_label`, `z_label` properties of the Axes handle are in fact Label handles.

Figure 23.9 : *Modifying coordinate axes labels*

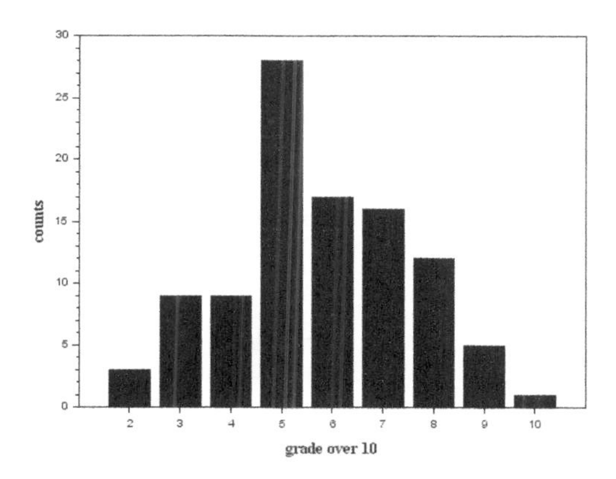

23.2. Creating animations

So far, we have covered how to generate a large number of different figure types, however all these figures were static. In this section, we will see how to create a graphical animation in Scilab. The basic idea is very simple: as for movies, you only need to link together several static figures to get an animation! You will therefore need to repeatedly call the graphics commands we have studied so far as well as use control flow structures such as loops as seen in Part *Programming*. We will start with a very simple animation: make a red dot of coordinates $x = \cos(2\pi t)$, $y = \sin(2\pi t)$ for $t = 0 \ldots 1$ move along its trajectory (the circle centered at $(0, 0)$ and of radius 1). We get the following program:

```
clf;
z=[0:0.01:2*%pi];  // to plot the trajectory
for t=0:0.01:1  // loop to increment the time
    // compute the new position
    x=cos(2*%pi*t); // x-coordinate
    y=sin(2*%pi*t); // y-coordinate
    clf; // parameterize the graphics window
    A=gca();A.data_bounds=[-1.2,-1.2;1.2,1.2];A.isoview="on";
    plot(cos(z),sin(z),'-b')   // trajectory in blue
    plot(x,y,'.r') // the animated dot in red
end
```

Let's analyze what this script does:

1. The `for` loop takes care of incrementing the values of `t` from 0 to 1, with a step of `0.01` (you can view `t` as the value of the elapsed time).
2. At each step, the point's coordinates `x,y` are recomputed as a function of `t`.
3. The graphics window is reset, its content cleared with `clf` and the plot size and scale are redefined *via* the `data_bounds` and `isoview` properties of the current Axes handle.
4. Then the circle (still the same) and the red dot (which changes position each time) are displayed again with `plot`.

Each step in the animation must display a red dot on a blue circle as shown in Figure 23.10.

Figure 23.10 : *A red dot moving along a blue circle (GIF)*

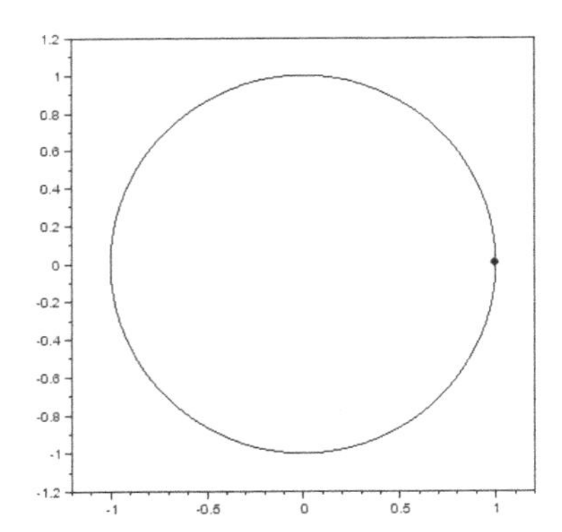

In fact, when you execute this script, the rendering is very poor! The animation is not smooth and the flashing is very disagreeable. This is normal since with the alternating `clf` and `plot` commands, the window is cleared for a short moment which gives the impression that the image is flashing. We will show you later how to get rid of this type of issue.

Exporting an animation

If you wish to export an animation, the easiest method consists in exporting each image that makes up the animation in `*.gif` format and then generate an animated GIF with an external tool. To export the different figures, use `xs2gif` right after the command(s) that display the plot that you wish to export (that would be the `plot` command in the previous script). Afterwards, if you know the ImageMagick [http://d-booker.jo.my/sci-imagemagick] software suite, you can easily create an animated GIF from a sequence of images stored inside `img_*.gif` files:

- by launching the following command line from a terminal:

```
convert img_*.gif -delay 10 -loop 0 animation.gif
```

- or straight from the Scilab console:

```
unix('convert img_*.gif -delay 20 -loop 0 animation.gif')
```

In practice, you need to modify the previous script by adding the command `xs2gif(0,'img_'+string(1000+k)+'.gif')` right after the `plot` command to create `gif` files inside the current directory.

```
clf;
k=0;  // to number the images
z=[0:0.01:2*%pi];  // to plot the trajectory
for t=0:0.01:1  // loop to increment the time
    // compute the new position
    x=cos(2*%pi*t); // x-coordinate
    y=sin(2*%pi*t); // y-coordinate
    clf; // parameterizing the graphics window
    A=gca();A.data_bounds=[-1.2,-1.2;1.2,1.2];A.isoview="on";
    plot(cos(z),sin(z),'-b')   // trajectory in blue
    plot(x,y,'.r') // the red point advances
    // to have time to see the figure:
    xs2gif(0,'img_'+string(1000+k)+'.gif')
    k=k+1  // increment the image number
end
unix('convert img_*.gif -delay 10 -loop 0 animation.gif')
mdelete('img_*.gif')
```

You thus need to add a variable `k` to your code to number the name of each figure temporarily. This numbering is performed by adding this number (here `1000+k`) to the file name as a string by using the command `string(1000+k)`. This means the first image is called `img_1000.gif` and the last `img_1100.gif` (and all the file names are the same size). You can then create the animated GIF with the `unix` command and delete all the temporary files with `mdelete`.

Improving the animation's smoothness

To avoid the flashing effect, use the commands `drawlater` and `drawnow` as shown in the following script.

```
clf;
z=[0:0.01:2*%pi];  // to plot the trajectory
for t=0:0.01:1  // loop to increment the time
    // compute the new position
    x=cos(2*%pi*t); // x-coordinate
    y=sin(2*%pi*t); // y-coordinate
    drawlater()
    clf; // parameterize the graphics window
    A=gca();A.data_bounds=[-1.2,-1.2;1.2,1.2];A.isoview="on";
    plot(cos(z),sin(z),'-b')   // trajectory in blue
    plot(x,y,'.r') // the red point advances
    drawnow()
    sleep(10) // to have enough time to see the figure
end
```

The animation is performed as if the result of the graphics commands located after `drawlater` is stored inside a graphics buffer and displayed only after the `drawnow` command. In fact, these two commands modify the `immediate_drawing` property of the current window's Figure handle, switching it from `"on"` to `"off"` and inversely.

Tip › *Often, the graphics command follow each other too fast to be able to visualize the changes inside the graphics window. In this case, add the command* **`sleep(10)`** *to force Scilab to add a pause (of ten thousandths of a second in this case), which will give you the time to see the graphics changes. We can also use the command* **`realtime`** *to control the times at which the displays are launched to get a simulation in real time (see the* ***script*** *in* ***Section 4.3, CPU dates and times****).*

You can also construct animations where the figure is not cleared at each step, for example by using `move` to relocate an entity in the graphics window. The previous animation can be rewritten as follows:

```
clf;
dt=0.01  // time step
z=[0:0.01:2*%pi];  // to plot the trajectory
A=gca();A.data_bounds=[-1.2,-1.2;1.2,1.2];A.isoview="on";
plot(cos(z),sin(z),'-b')   // trajectory in blue
plot(1,0,'.r') // initial position of the red dot
E=gce();  // handle of the red dot
for t=0:dt:1  // loop to increment the time
    dx=-2*%pi*sin(2*%pi*t)*dt; // displacement in x
    dy=2*%pi*cos(2*%pi*t)*dt; // displacement in y
    move(E,[dx,dy]);  // move the E handle
    sleep(10) // to have enough time to see the figure
end
```

The difficulty here lies in computing the displacement `dx,dy` to perform at each time step `dt` as a function of the trajectory you wish to follow. Here, a few calculations are needed to get:

$$dx \approx x'(t) \times dt = -2\pi \times \sin(2\pi t) \times dt, \; dy \approx y'(t) \times dt = 2\pi \times \sin(2\pi t) \times dt$$

After this, there is no need to clear the window and replot the circle: you no longer need the commands drawlater and drawnow to avoid the flashing effect. Here is another example that lets you avoid having to clear the window: if you wish to modify the appearance of a fixed window, you only need to modify the values of the current Axes handle, between drawlater and drawnow, to get an animation. Therefore, in order to rotate a three-dimensional figure, you only need to change the values of rotation_angles as shown in the following script.

```
clf;
plot3d1() // a three-dimensional surface
for k=1:360  // loop to rotate the figure
    // change the angle alpha of the view point by 1 degree
    A=gca();A.rotation_angles(2)=A.rotation_angles(2)+1;
    sleep(10) // to have enough time to see the figure
end
```

You will notice that rotating the figure about the z-axis is equivalent to you turning around the figure (see Figure 21.1 for an explanation of the angles that define the view point).

Figure 23.11 : Rotating a figure (GIF)

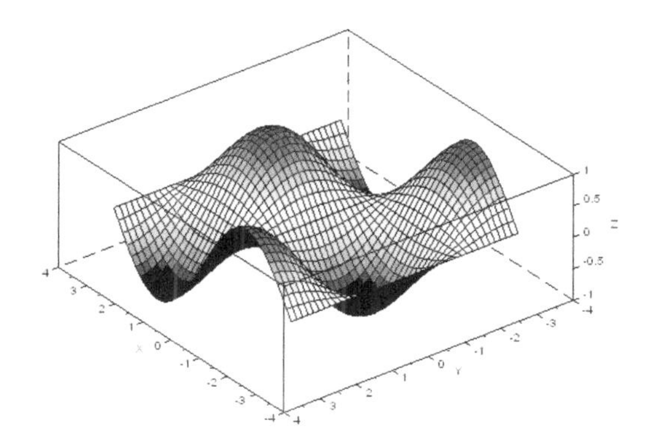

*Caution › The **drawlater** and **drawnow** commands act on the Figure handle of the current window. Be careful not to change the current figure between the **drawlater** and **drawnow** commands.*

Generating an animation without using loops

Certain simple animations can be created without having to use loops. This can be achieved by calling certain specific functions. For example comet and comet3d display a curve defined by the vectors x,y (and z for comet3d) by constructing it progressively. The plotting sequence is split up in three parts:

- the head, located at the coordinates x(i),y(i) and denoted by a little circle (marker 9 or o [see Figure 20.14 and Figure 20.2])
- it it followed by a tail, made up of a line of thickness 3 drawn over p percent of the length of the curve (p=0.1 by default)
- finally the body, meaning the rest of the curve, is displayed with a line of thickness 1.

You can plot several curves at the same time as long as they all have the same number of points and the same vectors x (and y in three dimensions). Try it with the following script, which yields the animations in Figure 23.12.

```
x=[1:1000]';
clf;
comet(x,[x zeros(x) -x],"colors",[2 3 5])
```

Figure 23.12 : *Animated curves with comet (GIF)*

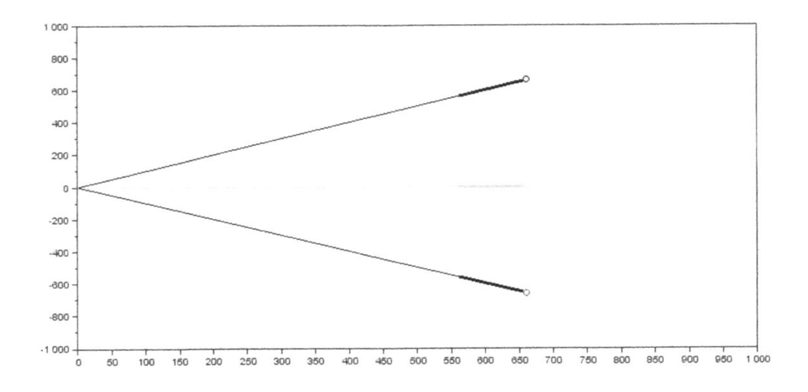

You can create the same animation by using the plot, drawlater and drawnow commands, but this involves more work:

```
x=[1:1000]';
// length of the tail as a percentage of the curve
p=0.1;
k=round(p*length(x));
```

```
// initialization of graphics window
clf;
drawlater()
plot([x x x],[x zeros(x) -x])
A=gca();rect=A.data_bounds;
for i=1:10:length(x)
    imk=max(1,i-k);
    drawlater()
    clf;
    // 1st curve in blue
    y=x;
    plot(x(1:i),y(1:i),'-b')  //  body
    plot(x(imk:i),y(imk:i),'-b')  // tail
    E=gce();E.children.thickness=3;
    plot(x(i),y(i),'ob')  // head
    // second curve in green
    y=zeros(x);
    plot(x(1:i),y(1:i),'-g')  // body
    plot(x(imk:i),y(imk:i),'-g')  // tail
    E=gce();E.children.thickness=3;
    plot(x(i),y(i),'og')  // head
    // third curve in red
    y=-x;
    plot(x(1:i),y(1:i),'-r')  // body
    plot(x(imk:i),y(imk:i),'-r')  // tail
    E=gce();E.children.thickness=3;
    plot(x(i),y(i),'or')  // head
    A=gca();A.data_bounds=rect;
    drawnow()
    sleep(10)
end
```

The syntax of `comet3d` is similar to that of `comet`. You can test this command with the following example:

```
t=[0:0.01:10]';
clf;
comet3d(sin(t),cos(t),[t, cos(t) sin(t) ],"colors",[2 3 5])
```

Animating a plot

The `paramfplot2d` command lets you create an animated plot depending on a parameter. To do this, you need to create a function of two variables `f(x,t)`.

```
function y=f(x,t)
    y=exp(-(x-t).^2/2)
endfunction
x=[-10:0.05:10].';
t=[-5:0.2:5];
clf;
paramfplot2d(f,x,t);
```

In this example, x is the discretization of the interval over which we are plotting the function f, with a curve plotted for each t value.

Figure 23.13 : *Curves animated with* `paramfplot2d` *(GIF)*

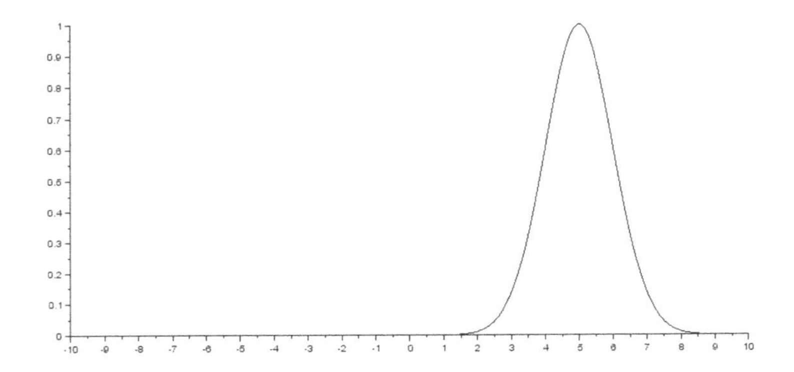

23.3. Interacting with the graphics window

In this section, you will learn how to interact with the graphics window with your mouse.

Retrieving a point's coordinates

You can retrieve the coordinates of a point selected with a mouse click by using the `locate` command. This command takes as input the number n of points which will be clicked, and returns a matrix with n columns and two rows with each column containing the coordinates of the points clicked. Execute the following script:

```
clf;
plot2d(0,0,0,rect=[0,0,1,1])
xgrid(4)
// click inside the graphics window three times
X=locate(3)
// plot the corresponding polygon
xfpolys(X(1,:)',X(2,:)',5)
// display the coordinates of the 5 points
txt=['List of the triangle''s three points:';'';
'P_'+string([1:3]')+'=('+string(X(1,:)')+','+string(X(2,:)')+')']
xstring(0.5,0.1,txt)
E=gce();
E.font_size=5;
E.font_style=3;
```

Once the script is loaded, click inside the window three times. The `locate` command retrieves the coordinates of the three points, then the rest of the script draws the corresponding triangle and displays the coordinates of the polygon's points in the graphics window as shown in Figure 23.14.

Figure 23.14 : *Retrieving coordinates with your mouse by using* `locate`

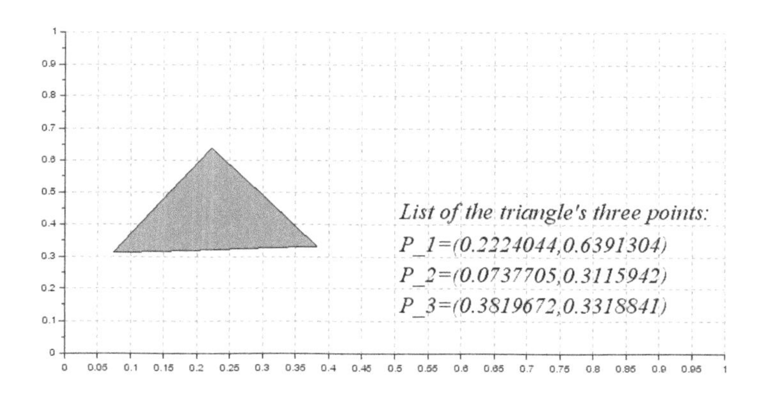

Differentiating different types of clicks

The `locate` command does not account for the type of click performed: single-click, double-click, right-click, left-click or center-click. If you wish to differentiate them, use the `xclick` command. This command not only lets you retrieve the coordinates of a click in the graphics window but also the type of click and which window is clicked. Launch the following script then click in the graphics window that opens up.

```
clf;
plot2d(0,0,0,rect=[0,0,1,1])
xgrid(4)
// wait for a click inside the window
[button,x,y,fig_id,menu]=xclick()
//  place a cross at the selected point
plot(x,y,'xk');
// display the point's coordinates
xstring(x,y,'  it''s point ('+string(x)+','+string(y)+')')
```

A cross and some text will appear in the graphics window at the location of the point clicked as in Figure 23.15. If you look at the data displayed in the console, you will notice that the click type was also retrieved by `xclick`.

```
-->clf;

-->plot2d(0,0,0,rect=[0,0,1,1])

-->xgrid(4)

-->// wait for a click inside the window

-->[button,x,y,fig_id,menu]=xclick()
 menu  =
 void
 fig_id  =
    0.
 y  =
    0.3086957
 x  =
    0.3469945
 button  =
    3.

-->//  place a cross at the selected point

-->plot(x,y,'xk');

-->// display the point's coordinates

-->xstring(x,y,'  it''s point ('+string(x)+','+string(y)+')')
```

Here we performed a simple click with the left button (`button=3`) inside figure 0. To know the value associated to each click type, enter `help xclick` in the console.

Figure 23.15 : *Retrieving coordinates with your mouse with* `xclick`

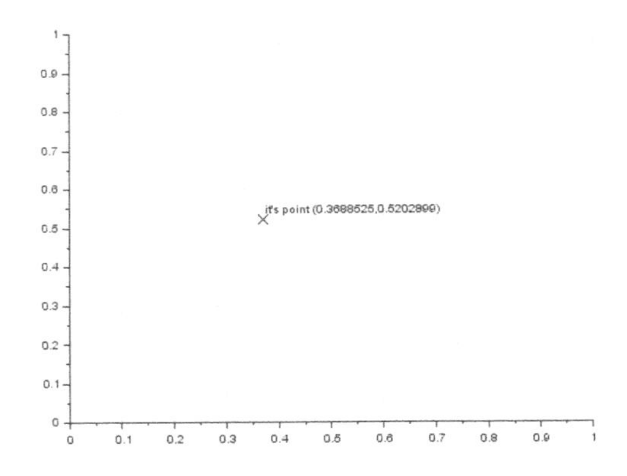

Retrieving all moves and clicks

You can retrieve all the mouse moves and clicks with `xgetmouse` and make the graphics window react accordingly. The `xgetmouse` command takes two boolean values as inputs:

- `getmotion` which must be set to `%f` to activate the movement detection
- `getrelease` which must be set to `%f` to activate click detection

According to these parameters, the command returns a vector of three values:

- The first two are the coordinates of the point in the graphics window.
- The last one indicates the state of the mouse according to the following codes (listed in the help page `help xgetmouse`).

In the following script, we will use `xgetmouse` to display (inside the status bar) the current location of the mouse inside the graphics window until the window gets clicked.

```
function [x,y]=coordinates()
    rep=-ones(1,3)
    // move the mouse until there's a click
    while rep(3)==-1
        rep=xgetmouse([%t %t])
        x=rep(1);y=rep(2);
        // display the coordinates
        txt='('+string(x)+','+string(y)+')'
        xinfo(txt)
    end
    // display the point in the graphics window as output
    plot(x,y,'xk');
    xstring(x,y,'  this is point ('+string(x)+','+string(y)+')')
endfunction

clf;
plot2d(0,0,0,rect=[0,0,1,1])
xgrid(4)
[x,y]=coordinates()
```

After clicking, the point's position appears in the window and the coordinates are output in the console. You can then regain control of the console.

Figure 23.16 : Retrieving coordinates with your mouse with `xgetmouse`

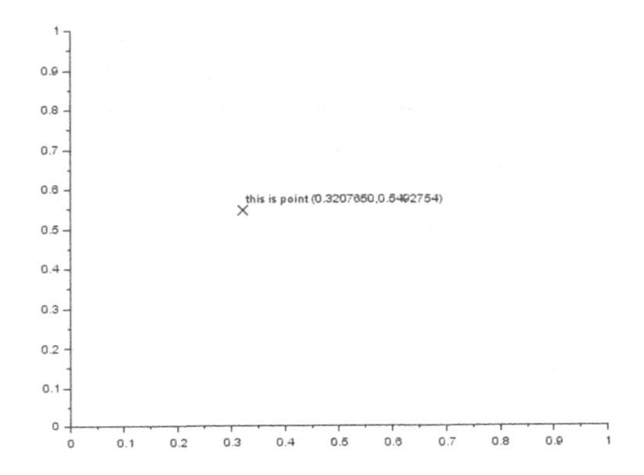

Tip › As seen here, you can use the **`xinfo`** *command to display a string of characters in the status bar at the bottom of the graphics window:*

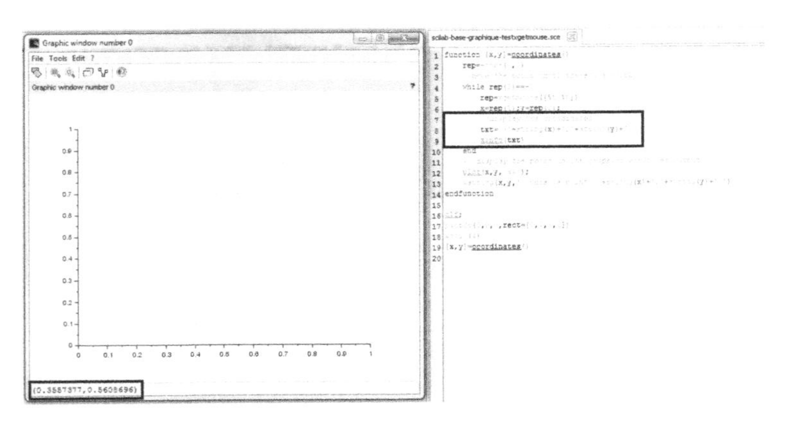

Parameterizing your own functions to retrieve events linked to the mouse

To go even further and parameterize your own functions that can retrieve events linked to the mouse or the keyboard, you will need to use `seteventhandler`. For example:

```
function mouse_action(win, x, y, iaim)
```

```
     if iaim==-1000 then return,end
     // convert coordinates from pixels to numbers
    [x,y]=xchange(x,y,'i2f')
    if iaim==-1 then // display coordinates
        txt='coordinates ('+string(x)+','+string(y)...
        +') window '+string(win)+' button '+string(iaim)
        xinfo(txt)
    else xstring(x,y,'you have clicked the mouse ('+...
        string(iaim)+') on ('+string(x)+','+string(y)+')')
    end
endfunction
// launching the graphics window
clf;
plot2d(0,0,0,rect=[0,0,1,1])
xgrid(4)
// activate the function mouse_action
seteventhandler('mouse_action')
// to stop the behavior of the graphics window
seteventhandler('')
```

Once the script is loaded, if you look at the graphics window, you will continuously see a message displayed in the status bar. This message follows the format `coordinates (0.4325425,0.2349875) window 0 button -1` and at each click, the text displays the position and value of `iaim` in the graphics window as shown in Figure 23.17. This behavior for the graphics window is sustained up until the window is closed or the user enters `seteventhandler('')` in the console. In fact, the command `seteventhandler` acts on the Figure handle's properties as follows:

- `event_handler` contains the name of the function that tracks the events (here `mouse_action`).
- `event_handler_enable` activates or deactivates event tracking according to its value (set at `"on"` or `"off"`).

For more details, see the corresponding help page by entering `help 'event handler functions'` in the console.

***Figure 23.17 :** Tracking graphics events with* `seteventhandler`

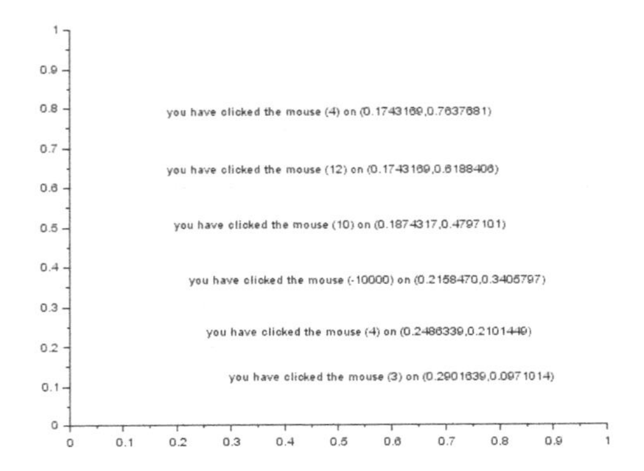

Caution › *The coordinates retrieved with the function* **`mouse_action`** *are provided as pixels rather than coordinates. To convert this, use the* **`xchange`** *command with the option* `"i2f"` *(option* `"f2i"` *for the opposite conversion).*

23.4. Creating your own graphical interfaces

A graphical user interface (often called GUI) is a graphics window that lets you perform a variety of operations without having to go through the console and enter commands. Some examples of operations are:

- launching a computation *via* a button
- choosing an option by checking a box
- entering a string of characters or a numerical value inside an entry field

Parameterizing a graphical interface window

Scilab offers the ability to create such windows. The interface generated is contained inside a Scilab graphics window and, therefore, is associated to a Figure handle. To create a graphical interface, start by creating a new graphics window by calling the command `figure`, as shown in the following script.

```
hfig = figure(...
  'Tag', 'myfigure', ...
  'BackgroundColor', name2rgb("lightblue")/255, ...
```

```
    'Figure_name', 'My Figure', ...
    'Position', [20 20 300 200])
```

The arguments of the `figure` command define the main properties of the figure in the form of a list of instructions such as `'property',value,etc`. In this case, we only set the size, color and name of the figure along with the `Tag` property which will be further used later on. Execute this script to get the result in Figure 23.18.

Figure 23.18 : *Creating a graphics window with* `figure`

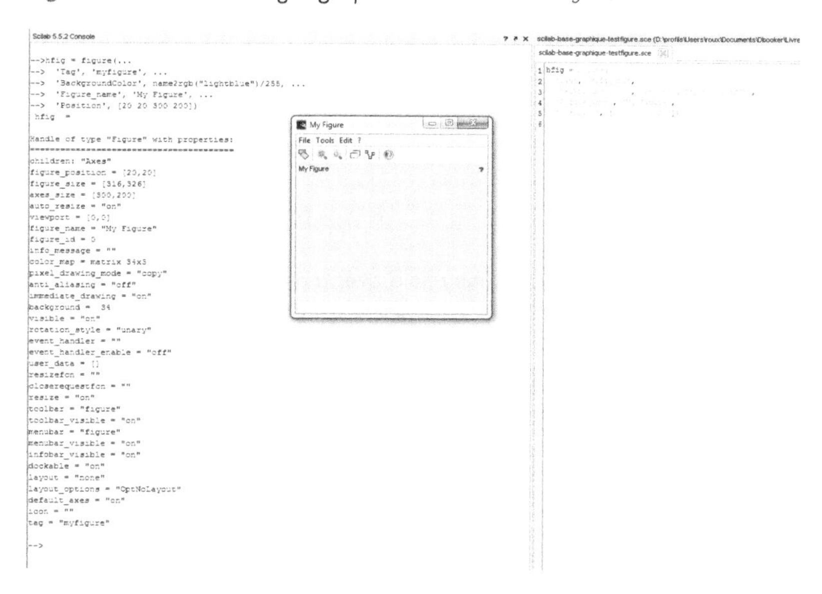

*Caution › The color options in the **`figure`** commands must be provided in RGB level format with values between 0 and 1. If you wish to use the commands **`name2rgb`** to designate a color by their name, you will need to divide the result by 255 to obtain levels between 0 and 1!*

The main elements of a graphical user interface

The elements that make up a graphical user interface will be linked to a new type of handle: the *uicontrols*. To create them, you will have to use the command `uicontrol`. Its syntax is very similar to that of the `figure` command:

1. Indicate the handle of the figure in which the uicontrol gets activated.
2. Define the values of the uicontrol's properties one after the other: `...,'property',value,...`.

To start, look at the following script to see how to add a simple text to your interface.

```
hfig = figure(...
  'Tag', 'myfigure', ...
  'BackgroundColor', name2rgb("lightblue")/255, ...
  'Figure_name', 'My Figure', ...
  'Position', [20 20 300 200]);

hobj = uicontrol(hfig, ...
  'Style', 'edit', ...
  'Tag', 'val', ...
  'String', 'enter text', ...
  'HorizontalAlignment', 'left', ...
  'FontAngle', 'italic', ...
  'FontSize', 20, ...
  'FontWeight', 'bold', ...
  'BackgroundColor', name2rgb("gray")/255, ...
  'ForegroundColor', name2rgb("black")/255, ...
  'FontUnits', 'pixels', ...
  'FontName', 'helvetica', ...
  'Position', [50 100 200 20]);
```

The `uicontrol` command includes a text entry field over a gray background as shown in Figure 23.19, with a black text font. The `Style` property determines the uicontrol type that gets generated. Here the type is *edit* which means it applies to an editable zone. You can then perform anything you are accustomed to doing with an entry field: erase the default text `enter text`, copy, paste a text previously copied, etc.

Figure 23.19 : *Creating a uicontrol*

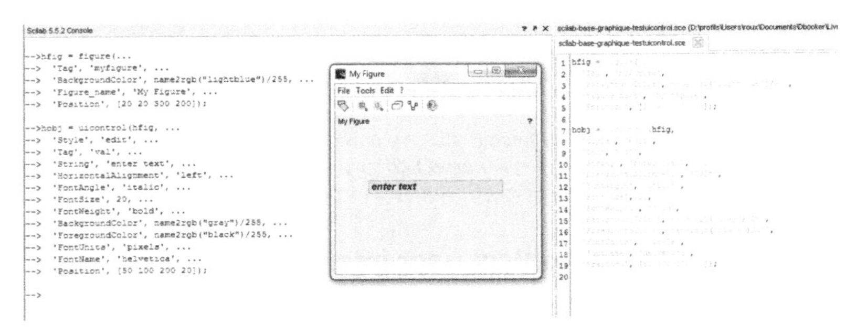

Tip › The number of properties that need to be defined is often very large. To improve the code legibility, it is preferable to enter only one property per line. To avoid syntax errors while loading the script, use `...` at the end of each line.

*Caution › The Uicontrol handles come directly from a Figure handle and not an Axes handle. The coordinates of the **`position`** properties are expressed as pixels with respect to the size of the current figure and not with respect to a partial scale associated to an Axes handle.*

There exist a lot of types of uicontrols according to the value of the `Style` property. You will find a complete list by entering `help uicontrol` inside the console. Here are several common examples:

- `text` to display a string of characters (which are given by the `value` property of the corresponding handle)
- `edit` to display an entry field (the contents of the `value` property of the corresponding handle, by default, gets displayed in this entry field)
- `pushbutton` to add a clickable button (whose display name will be the content of the `value` property of the corresponding handle)
- `checkbox` to add a checkbox or `radiobutton` (the content of the `value` property indicates if the box is checked or no, and the contents of the `string` field gets displayed next to the checkbox or button)
- `slider` to modify the value with the help of the slider

You can try out these different elements on the examples of the following script which produces the graphical user interface in Figure 23.20.

```
hfig = figure(...
  'Tag', 'myfigure', ...
  'BackgroundColor', name2rgb("lightblue")/255, ...
  'Figure_name', 'My Figure', ...
  'Position', [20 20 300 250]);

hobj1 = uicontrol(hfig, ...
  'Style', 'text', ...
  'Tag', 'val1', ...
  'String', 'a text', ...
  'HorizontalAlignment', 'left', ...
  'FontAngle', 'italic', ...
  'FontSize', 20, ...
  'FontWeight', 'bold', ...
  'BackgroundColor', name2rgb("gray")/255, ...
  'ForegroundColor', name2rgb("black")/255, ...
  'FontUnits', 'pixels', ...
  'FontName', 'helvetica', ...
  'Position', [50 200 50 20]);


hobj2 = uicontrol(hfig, ...
  'Style', 'edit', ...
  'Tag', 'val2', ...
  'String', 'entry field', ...
  'HorizontalAlignment', 'left', ...
  'FontAngle', 'italic', ...
  'FontSize', 20, ...
  'FontWeight', 'bold', ...
```

```
      'BackgroundColor', name2rgb("gray")/255, ...
      'ForegroundColor', name2rgb("black")/255, ...
      'FontUnits', 'pixels', ...
      'FontName', 'helvetica', ...
      'Position', [50 150 200 20]);

    hobj3 = uicontrol(hfig, ...
      'Style', 'slider', ...
      'Tag', 'val3', ...
      'String', 'slider', ...
      'HorizontalAlignment', 'left', ...
      'FontAngle', 'italic', ...
      'FontSize', 20, ...
      'FontWeight', 'bold', ...
      'BackgroundColor', name2rgb("gray")/255, ...
      'ForegroundColor', name2rgb("black")/255, ...
      'FontUnits', 'pixels', ...
      'FontName', 'helvetica', ...
      'Position', [50 100 200 20]);

    hobj4 = uicontrol(hfig, ...
      'Style', 'checkbox', ...
      'Tag', 'val4', ...
      'String', 'checkbox', ...
      'Value', 1, ...  // the box is checked!
      'HorizontalAlignment', 'left', ...
      'FontAngle', 'italic', ...
      'FontSize', 20, ...
      'FontWeight', 'bold', ...
      'BackgroundColor', name2rgb("gray")/255, ...
      'ForegroundColor', name2rgb("black")/255, ...
      'FontUnits', 'pixels', ...
      'FontName', 'helvetica', ...
      'Position', [50 50 200 20]);
```

Figure 23.20 : *Different types of uicontrols*

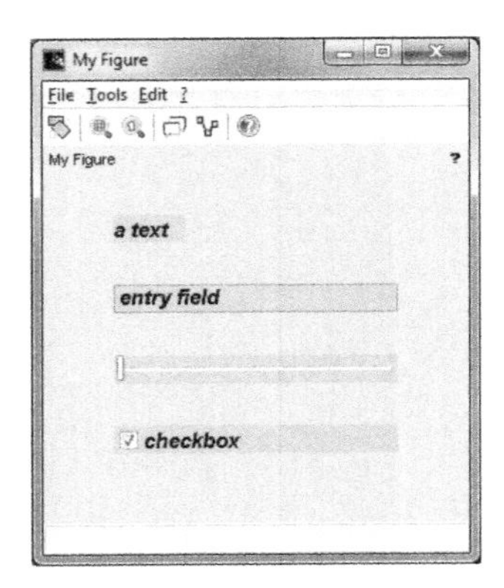

Attaching a task to an event

For the moment, these different elements do not have a use: you can change the entry field, slide the slider or uncheck the box... these actions don't have any effect on Scilab's behavior. To change this, you need to associate a callback function to one of the graphical interface's uicontrols so that the actions get converted into commands executed by Scilab. A callback is consequently a Scilab function which retrieves certain values associated to the different elements of the graphical interface, gets called by one of the uicontrols and eventually modifies the figure. To better understand this concept, execute the following script.

```
function myCallBack()
    // retrieve the handles called by their "Tag"
    // the first value: a
    h1=findobj("Tag","val1");
    // the second value: b
    h2=findobj("Tag","val2");
    // the result a+b
    h3=findobj("Tag","result");
    // compute the sum
    val1=h1.String;  // value of a
    val2=h2.String;  // value of b
    value=string(evstr(val1+'+'+val2))
    // assign the sum to the result box
    h3.string=value
endfunction


function myGui()
// MAIN WINDOW
hfig = figure(...
  'Tag', 'myGui', ...
  'BackgroundColor',  name2rgb("blue")/255, ...
  'Figure_name', 'myGui', ...
  'Position', [20 20 300 200]);

// 1st TEXT FIELD  (a)
h1 = uicontrol(hfig, ...
  'Style', 'edit', ...
  'Tag', 'val1', ...
  'String', '1', ...
  'HorizontalAlignment', 'left', ...
  'FontAngle', 'italic', ...
  'FontSize', 20, ...
  'FontWeight', 'bold', ...
  'BackgroundColor', name2rgb("gray")/255, ...
  'ForegroundColor', name2rgb("black")/255, ...
  'FontUnits', 'pixels', ...
  'FontName', 'helvetica', ...
  'Position', [50 100 50 20]);

// +
```

```
h4 = uicontrol(hfig, ...
  'Style', 'text', ...
  'Tag', 'plus', ...
  'String', '+', ...
  'HorizontalAlignment', 'center', ...
  'FontAngle', 'italic', ...
  'FontSize', 20, ...
  'FontWeight', 'bold', ...
  'BackgroundColor', name2rgb("gray")/255, ...
  'ForegroundColor', name2rgb("black")/255, ...
  'FontUnits', 'pixels', ...
  'FontName', 'helvetica', ...
  'Position', [100 100 20 20]);

// 2nd TEXT FIELD (b)
h2 = uicontrol(hfig, 'Style', 'edit', ...
  'Tag', 'val2', ...
  'String', '2', ...
  'HorizontalAlignment', 'left', ...
  'FontAngle', 'italic', ...
  'FontSize', 20, ...
  'FontWeight', 'bold', ...
  'BackgroundColor', name2rgb("gray")/255, ...
  'ForegroundColor', name2rgb("black")/255, ...
  'FontUnits', 'pixels', ...
  'FontName', 'helvetica', ...
  'Position', [120 100 50 20]);

// '='
h5 = uicontrol(hfig, ...
  'Style', 'text', ...
  'Tag', 'equal', ...
  'String', '=', ...
  'HorizontalAlignment', 'center', ...
  'FontAngle', 'italic', ...
  'FontSize', 20, ...
  'FontWeight', 'bold', ...
  'BackgroundColor', name2rgb("gray")/255, ...
  'ForegroundColor', name2rgb("black")/255, ...
  'FontUnits', 'pixels', ...
  'FontName', 'helvetica', ...
  'Position', [170 100 30 20]);

// 3rd TEXT FIELD  (result a+b)
h3 = uicontrol(hfig, ...
  'Style', 'edit', ...
  'Tag', 'result', ...
  'String', '', ...
  'HorizontalAlignment', 'left', ...
  'FontAngle', 'italic', ...
  'FontSize', 20, ...
  'FontWeight', 'bold', ...
  'BackgroundColor', name2rgb("gray")/255, ...
  'ForegroundColor', name2rgb("red")/255, ...
  'FontUnits', 'pixels', ...
  'FontName', 'helvetica', ...
```

```
  'Position', [200 100 50 20]);

// 'execute' BUTTON
buttonexec = uicontrol(hfig, ...
  'Style', 'pushbutton', ...
  'Tag', 'button1', ...
  'String', 'compute', ...
  'Callback','myCallBack()', ... // call the sum computation
  'BackgroundColor', name2rgb("green")/255, ...
  'ForegroundColor', name2rgb("black")/255, ...
  'FontAngle', 'normal', ...
  'FontSize', 20, ...
  'FontWeight', 'normal', ...
  'FontUnits', 'pixels', ...
  'FontName', 'helvetica', ...
  'Position', [50 50 100 20]);

// 'close' BUTTON
buttonclose = uicontrol(hfig, ...
  'Style', 'pushbutton', ...
  'Tag', 'button2', ...
  'String', 'close', ...
  'Callback','delete(findobj(''"Tag'",'"myGui'"))', ... // delete the
 window
  'BackgroundColor', name2rgb("green")/255, ...
  'ForegroundColor', name2rgb("black")/255, ...
  'FontAngle', 'normal', ...
  'FontSize', 20, ...
  'FontWeight', 'normal', ...
  'FontUnits', 'pixels', ...
  'FontName', 'helvetica', ...
  'Position', [160 50 100 20]);

endfunction

myGui()
```

You should get the graphical interface shown in Figure 23.21. This interface is comprised of seven uicontrols, which are created by calling the function `myGui()`:

- three handles `h1` , `h2` and `h3`, which are uicontrols of type *edit*, and the first two contain the strings `1` and `2`
- two handles `h4` and `h5`, which are uicontrols of type *text* that contain the strings `+` and `=`
- two handles `buttonexec` and `buttonclose` which are uicontrols of type *pushbutton* displayed in green and labeled Compute and Close

Once the graphics interface is launched, you can modify the contents of the editable zones `h1` and `h2` and enter different numerical values. When you push the Compute button, the sum of the numbers entered inside h1 and h2 gets displayed (in red) in zone h3 as shown in Figure 23.21. The function myCallback() computes the sum by retrieving the values contained inside the handles h1 and h2 and then reveals the result contained in the handle h3.

Figure 23.21 : A simple example of a graphical interface including callback

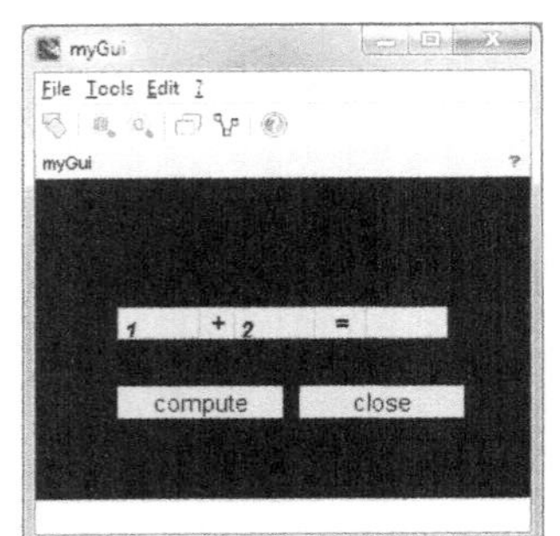

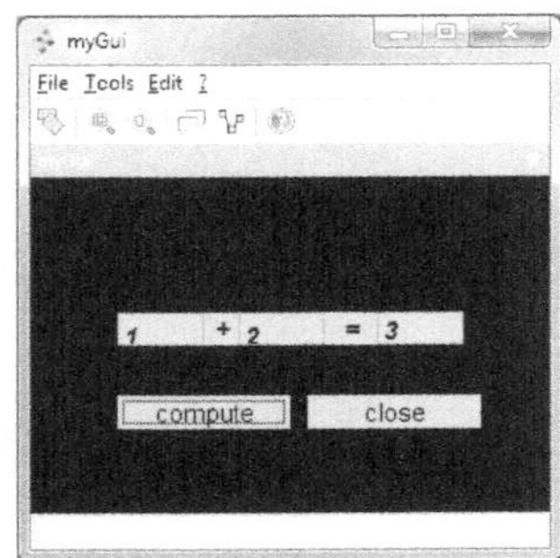

*Tip › Here it is very useful to add a value to the **Tag** property of each handle so as to retrieve the handle h with the command **findobj** by writing h=findobj("Tag","mytag").*

*Caution › We do not always have to explicitly create a function for each callback. In the example in **Figure 23.21**, if you press the Close button, the window closes. This is because the associated **buttonclose** performs a call to the command delete (findobj('"Tag'", '"myGui'")), in its **callback** field which deletes the figure with the **Tag** **myGui**. Here the callback is reduced to a fairly simple command and is therefore entered straight into the definition of uicontrol.*

Automatically refreshing the elements of a GUI

If you wish to repeatedly act on a graphical interface, the elements of the interface must be updated according to the different actions. These changes are performed by the repeated calls to the different callbacks associated with the interface elements. For this purpose, we locate them inside a loop that only gets exited once it is time to close the graphical interface. To explain this concept, let's look at the example of an interface which plots the function $f(x) = \cos(x)\ \cos(ax)$ over the interval $[-10;10]$ according to the value of a which we want to edit with the help of the slider.

Figure 23.22 : Plot a function of one parameter controlled by a slider (vidéo)

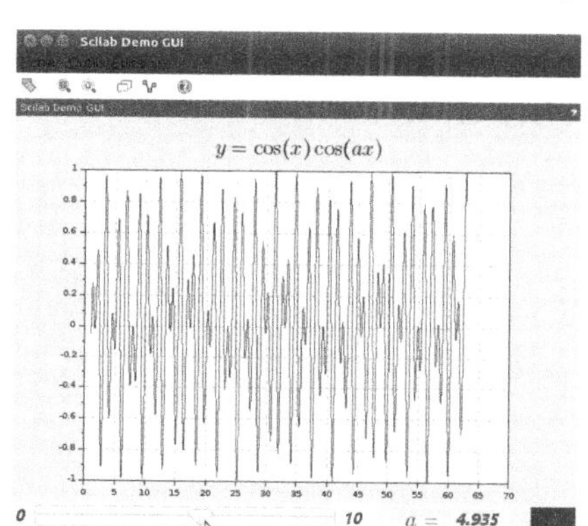

Analyzing this interface helps understand the different elements that need to be programmed. First, the graphical interface is represented by a main handle (let's call it `G`) whose children are the handles of the other elements:

- A handle `A` corresponds to the plot's curve.
- A handle `F`, uicontrol of type *frame*, contains the other interface elements:
  - The handle `S`, uicontrol of type *slider*, corresponds to the slider.
  - The handle `T`, uicontrol of type *text*, displays the value of the slider.
  - The handle `B`, uicontrol of type *pushbutton*, displays the button used to close the window.
  - Two other handles correspond to the values displayed on the left and right of the slider.

Among these handles, there are three elements that need to be updated according to the events performed. They correspond to the three callbacks that need to be defined:

- `plot_callback` to update the plot's appearance
- `disp_callback` to update the value of the slider
- `quit_callback` to close the interface

Here is the Scilab code that corresponds to all these elements:

```
//*********************************************
//  Setting up the interface
//*********************************************

Height=480;
Width=640;

// the figure
G = figure('position', [0 0 Width Height],...
'Tag','figure_handle',...
'backgroundcolor', name2rgb('lightblue')/255,...
"figure_name", 'Scilab Demo GUI');

//A: Axes that contain the graph
A=newaxes();
A.auto_clear="on"; // automatically clear

// F contains the uicontrols
F = uicontrol(G,...
'Tag','frame_handle',...
'position', [0 0 Width-2 35], ...
'fontsize', 12, ...
'style', 'frame',...
'string', 'buttons', ...
'backgroundcolor',  name2rgb('lightgray')/255);

// the slider
S = uicontrol('parent',F,...
'style', 'slider', ...
'Tag','slider_value',...
'position', [20 2 350 30],....
"Min",0,"Max",10,...
"value",0);

// several zones of text
T = uicontrol('parent',F,...
'style', 'text', ...
'Tag','slider_value_display',...
'position', [500 2  80 30],...
'String',' ',...
'callback','disp_callback', ...// displays the slider value
'backgroundcolor',  name2rgb('lightgray')/255,...
'HorizontalAlignment', 'left', ...
'FontAngle', 'italic', ...
'FontSize', 20, ...
'FontWeight', 'bold');

T1 = uicontrol('parent',F,...
'style', 'text', ...
'Tag','slider_value_min',...
'position', [2 2  18 30],...
'String', '0',...
'backgroundcolor',  name2rgb('lightgray')/255,...
```

```
'HorizontalAlignment', 'left', ...
'FontAngle', 'italic', ...
'FontSize', 20, ...
'FontWeight', 'bold');

T2 = uicontrol('parent',F,...
'style', 'text', ...
'Tag','slider_value_max',...
'position', [372 2  40 30],...
'String', '10',...
'HorizontalAlignment', 'left', ...
'FontAngle', 'italic', ...
'FontSize', 20, ...
'FontWeight', 'bold');

T3 = uicontrol('parent',F,...
'style', 'text', ...
'Tag','var_name',...
'position', [440 2  40 20],...
'String', '$$a=$$',...
'backgroundcolor',  name2rgb('lightgray')/255,...
'HorizontalAlignment', 'right', ...
'FontSize',20);

// a button to quit the interface
B = uicontrol('parent',F,...
'position', [Width-55,2 50 30], ...
'Tag','quit_button',...
'fontsize', 12, ...
'style', 'pushbutton',...
'string', 'Exit', ...
'callback', 'quit_callback', ...// will stop the loop and close the
 window
'backgroundcolor', name2rgb('red')/255);


//*********************************************
//  callbacks
//*********************************************

function disp_callback()
    // retrieve the slider value
    E1=findobj("Tag","slider_value");
    value=E1.value;
    // change the value displayed
    E2=findobj("Tag","slider_value_display");
    E2.string=string(value)
endfunction

function plot_callback()
    // retrieve the cursor value
    E1=findobj("Tag","slider_value");
    value=E1.value;
    // draw the curve
    drawlater()
    x=[1:0.001:20*%pi];
```

```
        y=cos(x).*cos(value*x)
        plot(x,y,'-r')
        xgrid(3)
        A=gca()
        A.title.text="$$y=\cos(x)\cos(a x)$$";
        A.title.font_size=4;
        drawnow()
    endfunction

    function quit_callback()
    h=findobj("Tag","figure_handle");// retrieve the window's handle
    delete(h);
    abort  // exit the while loop
    endfunction


    //*********************************************
    //  launch the interface
    //*********************************************

    cont=%t
    while cont // the main loop
        if findobj("Tag","figure_handle")==[] then
            cont=%f; // exit the window prematurely
            break
        else
            F.position=[0 0 G.figure_size(1)-2 35];
           disp_callback()  // update the value a
           plot_callback()  // update the plot
           end
    end
    delete(gcf())  // if quit the window without using the button B
```

The `while` loop at the end of the program refreshes the interface to display the new user updates by repeatedly calling the functions `disp_callback` and `plot_callback`. The command `F.position=[0 0 G.figure_size(1)-2 35];` modifies the position of the uicontrol `F` that contains the slider, the text zones and button as a function of the size of the window `G.figure_size(1)` (more precisely as a function of its width). This lets the graphics interface adjust to the changes in window size provided by the user.

Tip › *Gathering together the uicontrols inside a uicontrol of type* frame *lets you easily modify their placement in the window since you only need to change one handle to change the position of all its children. On the other hand you need to think of the elements' positions relative to the uicontrol of type* frame *rather than as absolute.*

Caution › *Different ways of exiting the graphical interface can lead to unrecoverable errors in the Scilab session. Specifically, if the graphics interface gets launched from the main execution level of Scilab, and if the figure gets prematurely closed, you can get errors such as:* The handle isn't or is no longer valid *or you may not even be able to stop the loop that refreshes the interface!*

Adding menus to the graphical user interface

Uicontrols also let you modify menus associated to a figure. To do this, you need to use the `uimenu` command which will create handles of type *uimenu* that are children of a Figure handle and will correspond to the new graphics window menus. The syntax is similar to that of the `uicontrol` command:

1. Specify the handle of the figure in which the menu gets activated.
2. Define the values of the uimenu one by one: `...,'property',value,...` specifically:
   - `label` for the menu name;
   - `callback` for the action associated to the menu.

Execute the following script to get the graphics interface in Figure 23.23.

```
hfig = figure(...
  'Tag', 'mafigure', ...
  'BackgroundColor', name2rgb("lightblue")/255, ...
  'Figure_name', 'My Figure', ...
  'Position', [20 20 300 200]);

hobj = uicontrol(hfig, ...
  'Style', 'text', ...
  'Tag', 'val', ...
  'String', 'adding menus', ...
  'HorizontalAlignment', 'left', ...
  'FontAngle', 'italic', ...
  'FontSize', 20, ...
  'FontWeight', 'bold', ...
  'BackgroundColor', name2rgb("gray")/255, ...
  'ForegroundColor', name2rgb("black")/255, ...
  'FontUnits', 'pixels', ...
  'FontName', 'helvetica', ...
  'Position', [50 100 200 20]);

//adding a menu
m=uimenu(hfig,'label', 'My Menu');
// item in menu m
m1=uimenu(m,'label', 'operations');
m2=uimenu(m,'label', 'delete the window', 'callback',
 "delete(findobj('"Tag'",'"myfigure'"))");
//sub-items ...
m11=uimenu(m1,'label', 'new window', 'callback',"figure(''Figure_name'',
 ''new'')");
m12=uimenu(m1,'label', 'clear window', 'callback',"clf()");
```

You will get the following in the menu bar:

- a menu called OPERATIONS that opens:
  - a sub-menu NEW WINDOW which opens a new graphics window
  - a sub-menu CLEAR WINDOW which clears the current graphics window
- a menu called DELETE WINDOW which closes the current window

The three actions NEW WINDOW, CLEAR WINDOW and DELETE WINDOW are associated to the commands written in the `callback` fields of the associated handles.

Figure 23.23 : *Creating a menu with* `uimenu`

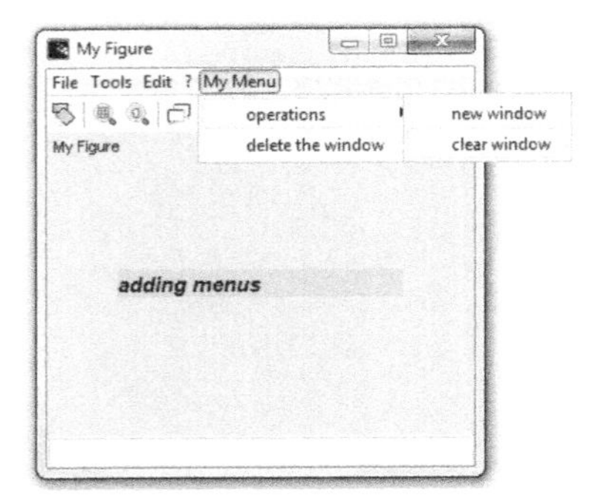

Tip › A sub-menu is just a menu that has a handle of type uimenu *as parents rather than a handle of type* figure. *It is very easy to add on sub-menus!*

Similarly, you can add a context menu to a graphics interface by using the command `uicontextmenu`. Execute the following script to get the context menu in Figure 23.24.

```
// creating a menu
hMenu = uicontextmenu();
// creating sub-menus
hItem1 = uimenu("Label", "Menu1", "Parent", hMenu, "Callback",
 "disp(""Menu1 selected!"")");
hItem2 = uimenu("Label", "Menu2", "Parent", hMenu, "Callback",
 "disp(""Menu2 selected!"")");
```

The `uicontextmenu` command creates a context menu in the window in which it is launched (in this case, in the console). The sub-menus defined by the Uimenu handles (created with the `uimenu` function) are children of the main menu. The actions associated to each menu are once again commands linked to the `callback` field of each menu.

Figure 23.24 : *Creating context menus with* `uicontextmenu`

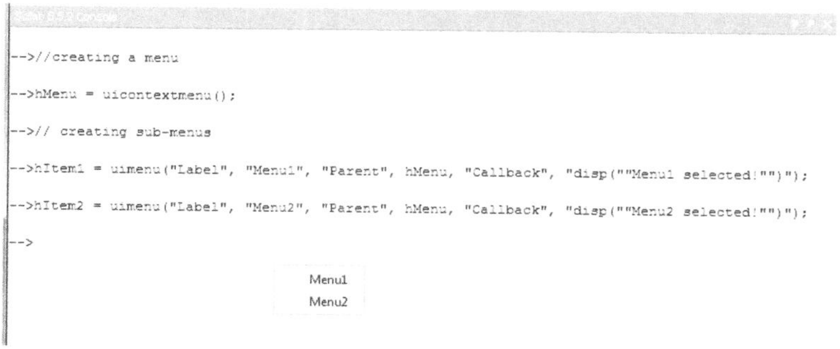

```
-->//creating a menu

-->hMenu = uicontextmenu();

-->// creating sub-menus

-->hItem1 = uimenu("Label", "Menu1", "Parent", hMenu, "Callback", "disp(""Menu1 selected!"")");

-->hItem2 = uimenu("Label", "Menu2", "Parent", hMenu, "Callback", "disp(""Menu2 selected!"")");

-->

                                        Menu1
                                        Menu2
```

Here the callbacks consist in only one `disp` command which displays a string in the console:

```
-->hMenu = uicontextmenu();

-->// creating sub-menus

-->hItem1 = uimenu("Label", "Menu1", "Parent", hMenu, "Callback",
 "disp(""Menu1 selected!"")");

-->hItem2 = uimenu("Label", "Menu2", "Parent", hMenu, "Callback",
 "disp(""Menu2 selected!"")");
```

***Caution* › *When you write a sequence of commands inside the* `callback` *fields of a uicontrol, what gets written are character strings. Make sure you close the quotation marks* ' *and* ".**

24

Two Case Studies: a Pendulum and Comet Orbit

In this last chapter, we are going to cover how to create the animations introduced in chapter *Preview of Scilab*. These animations involve all fundamental knowledge of Scilab:

- There are Scilab scripts that call or define several functions (see Part *Programming*).
- Among the different instructions, several directly affect the graphics window (refer back to Part *Creating Plots*).
- Finally, these scripts perform numerous computations that involve matrices and mathematical functions (which were covered in Part *Computing*).

For each animation, we will explain how the code was generated in order for you to adapt it to other problems.

Caution › To ensure the scripts introduced in this section are well executed, use Scilab version 5.4.1 or later.

24.1. The spring pendulum

The first animation simulates the movement of a pendulum attached to a spring as shown in Figure 24.1.

The animation is broken into two pieces:

1. The user interacts with the graphical user interface to choose the initial position of the pendulum with a click of the mouse.
2. The rest of the script is then in charge of displaying the resulting pendulum motion, following the laws of physics given by the following equations:

$$\begin{cases} r\times\frac{\mathrm{d}^2a}{\mathrm{d}t^2}+2\frac{\mathrm{d}r}{\mathrm{d}t}\times\frac{\mathrm{d}a}{\mathrm{d}t}=g\times\sin(a) \\ \frac{\mathrm{d}^2r}{\mathrm{d}t^2}-\frac{k}{m}(r-r_0)=r\times\left(\frac{\mathrm{d}a}{\mathrm{d}t}\right)^2+g\times\cos(a) \end{cases}$$

where :

- a is the pendulum angle with respect to the vertical
- r is the length of the spring that constitutes the pendulum

and the initial conditions are: $a(0)=a_0,\ \frac{\mathrm{d}a(0)}{\mathrm{d}t}=0,\ \ r(0)=r_0,\ \frac{\mathrm{d}r(0)}{\mathrm{d}t}=0.$

Figure 24.1 : *Simulation of a pendulum hung by a spring (GIF)*

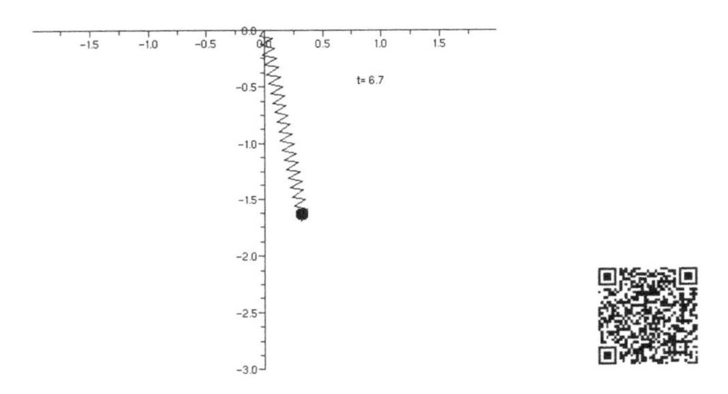

The script is fairly long so in order to understand it, we split it up into several steps.

```
//*****************************************************************
// animation of a spring pendulum
//*****************************************************************
//  function to create rotation matrix
function M=rot(a)
    M=[cos(a),sin(a);-sin(a),cos(a)];
endfunction
//  constants
n=40;      // number of coils of the spring
T=5;       // duration of the simulation
g=9.81;    //  g (gravitational acceleration)
k=10;      //  k (spring stiffness)
dt=0.01;   // dt (time step)

//*****************************************************************
// launch the graphics window
//*****************************************************************
//  window title
xtitle("(left-click to start the animation, right-click to stop)")
// title page (in LaTeX)
titlepage(["numerical solution of the spring pendulum ODE : ";" "; "$
$\Large r{d^2\over dt^2}a+2{d\over dt}r \times {d\over dt}a=g\times
 \sin(a)$$";
```

```
" "; "$$\Large {d^2\over dt^2}r-{k\over m}(r-r_0)=r\left({d\over dt}
 a\right)^2+g\times \cos(a)$$";" "; " with initial conditions : "; "$$
\Large  a(0)=? \;\;\;\;\;\; {d\over dt}a(0)=0  \;\;\;\;\;\; r(0)=r_0=?
 \;\;\;\;\;\; {d\over dt}r(0)=0 $$"])

//*****************************************************************
// processing the graphics window interactions
//*****************************************************************
// wait for a mouse click in the window
[c_i,c_x,c_y,c_w]=xclick();
while (c_i<>2)&(c_i<>5)   // as long as there is no right-click
      clf()    //clear the window
      //***********************************************************
      // retrieve the animation's initial data
      //***********************************************************
      // window title
      xtitle("(one click to initialize pendulum position a(0) and
 r(0) )")
      // parameterize the window's Axes handle
      plot(0,0,'.k');A=gca();A.x_location="origin";
      A.y_location="origin";A.auto_scale="off";A.isoview="on";
      A.data_bounds=[-1 -1; 1,0];xgrid(3)
      // retrieve the pendulum's initial position coordinates :
      [c_i,x,y,c_w]=xclick();
      // compute initial values :
      a=sign(x)*abs(atan(x/y));a0=a;da=0;  // compute angle a(0)
      l=sqrt(x^2+y^2);r=l;,dr=0;           //  compute r(0)
      // adapt the window's size to the pendulum's maximum size :
      A.data_bounds=[-1.5,-max(4*l,4);1.5,max(l,0.5)];
      //***********************************************************
      // loop creates the animation
      //***********************************************************
      for t=0:dt:T
           //******************************************************
           // compute new positions
           //******************************************************
           // solve the differential equation for a and r using
           // Euler's method
           dda=-(g*sin(a)+2*dr*da)/r;
           ddr=r*(da)^2-k*(r-l)+g*cos(a);
           da=da+dt*dda;
           dr=dr+dt*ddr;
           a=a+dt*da;
           r=r+dt*dr;
           // compute the spring's line representation
           springr=linspace(0,r,n)';        // the spring stretches
           // coordinates transverse to spring's axis -> /\/\/\
           springa=[0;(-1).^[0:n-3]';0]*(l/10);
           //rotate the spring's picture by the angle a
           x=[x;r*sin(a)];
           y=[y;-r*cos(a)];
           M=-rot(-a);
           N=[springr,springa]*M;
           springy=N(:,1);springx=N(:,2);
           //******************************************************
           // display the pendulum
```

```
        //*********************************************************
        drawlater()  // write to the graphics buffer
        clf()        // clear the window
        plot(springx,springy) //display the pendulum's spring
        xstring(0,0.1,["t=" string(t)]) // elapsed time
        // pendulum bob :
        xfarc(r*sin(a)-0.05,-r*cos(a)+0.05,0.1,0.1,0,360*64)
        // resize the graphics window
        A=gca();A.data_bounds=[-1.5,-max(4*l,4);1.5,max(l,0.5)];
        A.auto_scale="off";A.isoview="on";
        A.axes_visible=["off" "off" "off"];
        drawnow()              // display the graphics buffer
        realtime(t);            // delay display
    end
    //*************************************************************
    // choose a new animation or exit program
    //*************************************************************
    xtitle("(one clic to proceed )")    // window title
    plot(x,y,'-r')                       // display trajectory
    A=gca();A.isoview="on";xgrid(3); // display grid (green)
    // waiting for a mouse click in graphics window :
    [c_i,x,y,c_w]=xclick();
    clf();                          // choose a new operation
    xtitle("(left-click to start the animations, right-click to
 stop)")
    plot(0,0,'.k');A=gca();A.x_location="origin";
    A.y_location="origin";
    // waiting for a mouse click in the window :
    [c_i,x,y,c_w]=xclick();
end
```

Tip › *Using comments in your long script codes helps make it easier to understand, by clearly separating each section and explaining the purpose of each command.*

To clearly understand this code, let's look at each section of the script one by one.

1. At the beginning of the script, we define a function `rot` which gets called at each step of the animation to compute the pendulum's rotation. We also define several constants used at each step in the computation. These constants appear in the pendulum's equations:

```
// function to create rotation matrix
function M=rot(a)
    M=[cos(a),sin(a);-sin(a),cos(a)];
endfunction
// constants
n=40;       // number of coils of the spring
T=5;        // duration of the simulation
g=9.81;     // g (gravitational acceleration)
k=10;       // k (spring stiffness)
dt=0.01;    // dt (time step)
```

2. After these initial setups, we display a title page which opens up a graphics window and guide the user through the simulation:

```
// window title
xtitle("(left-click to start the animation, right-click to stop)")
// title page (in LaTeX)
titlepage(["numerical solution of the spring pendulum ODE : ";" ";
"$$\Large r{d^2\over dt^2}a+2{d\over dt}r \times {d\over dt}a=g
\times \sin(a)$$";
" "; "$$\Large {d^2\over dt^2}r-{k\over m}(r-r_0)=r\left({d\over dt}
 a\right)^2+g\times \cos(a)$$";
" "; " with initial conditions : "; "$$\Large  a(0)=? \;\;\;\;\;\;
 {d\over dt}a(0)=0  \;\;\;\;\;\; r(0)=r_0=? \;\;\;\;\;\; {d\over
 dt}r(0)=0 $$"])
```

3. Next, the `while` loop deals with the user's selections, by tracking the mouse clicks performed inside the graphics window:
   - The type of click performed gets retrieved with the `xclick` command inside the `c_i` variable.
   - The `while` loop relaunches the animation as long as the right button is not pressed (`c_i=2`) or clicked (`c_i=5`) (read the documentation by entering `help xclick`).

```
// wait for a mouse click in the window
[c_i,c_x,c_y,c_w]=xclick();
while (c_i<>2)&(c_i<>5)   // as long as there is no right-click
while (c_i<>2)&(c_i<>5)
      clf()    // clear the window
      // 1) retrieve the initial data
      // 2) create the animation
      // 3) choose a new animation or exit the program
// waiting for a mouse click in the window:
      [c_i,x,y,c_w]=xclick();
end
```

There are therefore three steps that manage the animation inside this loop:

- It retrieves the initial data, which is performed with `xclick` after parameterizing the graphics window with `plot` and `gca`. It then computes the motion's initial values `a0` and `r0` as a function of the initial position coordinates:

```
// window title
xtitle("(one click to initialize pendulum position a(0) and
 r(0) )")
// parameterize the window's Axes handle
plot(0,0,'.k');A=gca();A.x_location="origin";
A.y_location="origin";A.auto_scale="off";A.isoview="on";
```

```
A.data_bounds=[-1 -1; 1,0];xgrid(3)
// retrieve the pendulum's initial position coordinates :
[c_i,x,y,c_w]=xclick();
// compute initial values :
a=sign(x)*abs(atan(x/y));a0=a;da=0;  // compute angle a(0)
l=sqrt(x^2+y^2);r=l;,dr=0;           // compute r(0)
// adapt the window's size to the pendulum's maximum size :
A.data_bounds=[-1.5,-max(4*l,4);1.5,max(l,0.5)];
```

- Creating the animation consists in a simple `for` loop which increments the time parameter `t` by a step `dt` up to the maximum animation time `T` (defined at the beginning of the script):

```
for t=0:dt:T
    // 1) Compute the new position
    // 2) display the new position
end
```

The main elements of the animation are located inside this loop:

Computation of the pendulum's new position

This computation is based on mathematical techniques (Euler's method) which are applied to the pendulum's equations previously provided.

```
// solve the differential equation for a and r using
// Euler's method
dda=-(g*sin(a)+2*dr*da)/r;
ddr=r*(da)^2-k*(r-l)+g*cos(a);
da=da+dt*dda;
dr=dr+dt*ddr;
a=a+dt*da;
r=r+dt*dr;
// compute the spring's line representation
springr=linspace(0,r,n)';        // the spring stretches
// coordinates transverse to spring's axis -> /\/\/\
springa=[0;(-1).^[0:n-3]';0]*(1/10);
// rotate the spring's picture by the angle a
x=[x;r*sin(a)];
y=[y;-r*cos(a)];
M=-rot(-a);
N=[springr,springa]*M;
springy=N(:,1);springx=N(:,2);
```

Display the pendulum

The spring is displayed as a broken line by using `plot`, and the pendulum bob is represented by a disk created with the command `xfarc`. These instructions are located between the `drawlater()` and `drawlnow()` commands to improve the animation's smoothness. The `sleep(10)` instruction lets you adjust the animation's rate according to the time step (`dt=0.01` seconds, i.e. `10` thousands of a second).

```
drawlater()  // write to the graphics buffer
clf()        // clear the window
plot(springx,springy) // display the pendulum's spring
xstring(0,0.1,["t=" string(t)]) // elapsed time
// pendulum bob :
xfarc(r*sin(a)-0.05,-r*cos(a)+0.05,0.1,0.1,0,360*64)
// resize the graphics window
A=gca();A.data_bounds=[-1.5,-max(4*l,4);1.5,max(l,0.5)];
A.auto_scale="off";A.isoview="on";
A.axes_visible=["off" "off" "off"];
drawnow()          // display the graphics buffer
realtime(t);         // delay display
end
```

- At the end, the user has the option of creating a new animation or exiting the script. This is again achieved with the help of `xclick`, `plot`, `gca`.

```
xtitle("(one clic to proceed )")    // window title
plot(x,y,'-r')                       // display trajectory
A=gca();A.isoview="on";xgrid(3); // display grid (green)
// waiting for a mouse click in graphics window :
[c_i,x,y,c_w]=xclick();
clf();                          // choose a new operation
xtitle("(left-click to start the animations, right-click to
 stop)")
plot(0,0,'.k');A=gca();A.x_location="origin";
A.y_location="origin";
// waiting for a mouse click in the window :
[c_i,x,y,c_w]=xclick();
```

24.2. Simulating a comet's orbit

The second animation simulates the trajectory of a comet orbiting around a star while perturbed by the presence of a planet.

Figure 24.2 : Simulation of a comet's orbit (GIF)

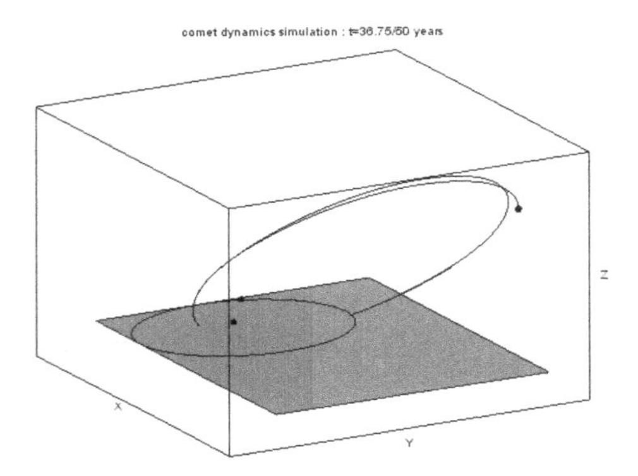

This is a particular case of the three-body problem, where the following simplifying assumptions are made:

- We choose a coordinate system where the origin coincides with the star's location. The star is considered fixed at the position $(0,0,0)$.
- We also assume the planet rotates around the star in the $z=0$ plane (to achieve this, its initial velocity must be constrained to this plane).
- Finally, we neglect the force or the comet acting on the plane (it is therefore only acting upon by the star) but we account for the effect of the planet on the comet (its movement is consequently perturbed).

If we define the positions of the different bodies with vectors of three coordinates:

$R_1 = \begin{pmatrix} x_1 \\ y_1 \\ z_1 \end{pmatrix}$ for the planet, $\quad R_2 = \begin{pmatrix} x_2 \\ y_2 \\ z_2 \end{pmatrix}$ for the comet.

The equations of motion are therefore written in the form of a system of 2nd order ordinary differential equations:

$$\begin{aligned} m_1\frac{d^2}{dt^2}R_1(t) &= -\frac{m_1 m_0}{|R_1(t)|^3}R_1(t) \\ m_2\frac{d^2}{dt^2}R_2(t) &= -\frac{m_2 m_0}{|R_2(t)|^3}R_2(t) - \frac{m_2 m_1}{|R_2(t)-R_1(t)|^3}(R_2(t)-R_1(t)) \end{aligned}$$

While defining the velocities of the planet and comet as

$$V_1 = \frac{d}{dt}R_1 = \begin{pmatrix} vx_1 \\ vy_1 \\ vz_1 \end{pmatrix} \text{ for the planet,} \qquad V_2 = \frac{d}{dt}R_2 = \begin{pmatrix} vx_2 \\ vy_2 \\ vz_2 \end{pmatrix} \text{ for the comet,}$$

We can rewrite this system of equations as a matrix equation of the form $\frac{d}{dt}u = F(t, u)$ where :

$$u = \begin{vmatrix} x_1 \\ y_1 \\ z_1 \\ vx_1 \\ vy_1 \\ vz_1 \\ x_2 \\ y_2 \\ z_2 \\ vx_2 \\ vy_2 \\ vz_2 \end{vmatrix} \quad \text{and} \quad F(t, u) = \begin{vmatrix} vx_1 \\ vy_1 \\ vz_1 \\ -\frac{m_0}{|R_1(t)|^3}x_1(t) \\ -\frac{m_0}{|R_1(t)|^3}y_1(t) \\ -\frac{m_0}{|R_1(t)|^3}z_1(t) \\ vx_2 \\ vy_2 \\ vz_2 \\ -\frac{m_0}{|R_1(t)|^3}x_2(t) - \frac{m_1}{|R_2(t)-R_1(t)|^3}(x_2(t)-x_1(t)) \\ -\frac{m_0}{|R_1(t)|^3}y_2(t) - \frac{m_1}{|R_2(t)-R_1(t)|^3}(y_2(t)-y_1(t)) \\ -\frac{m_0}{|R_1(t)|^3}z_2(t) - \frac{m_1}{|R_2(t)-R_1(t)|^3}(z_2(t)-z_1(t)) \end{vmatrix}$$

The advantage of using this formulation resides in the fact that Scilab lets you solve this type of equation numerically with the help of the `ode` command. You can find documentation on the `ode` function by entering these commands in the Scilab console:

- `help ode` if you already know the name of the command

- `apropos 'differential equation'` if you have no idea what command can be used to solve differential equations

In both cases, you should get redirected to the correct help page as shown in Figure 24.3.

***Figure 24.3 :** Help page for ode function*

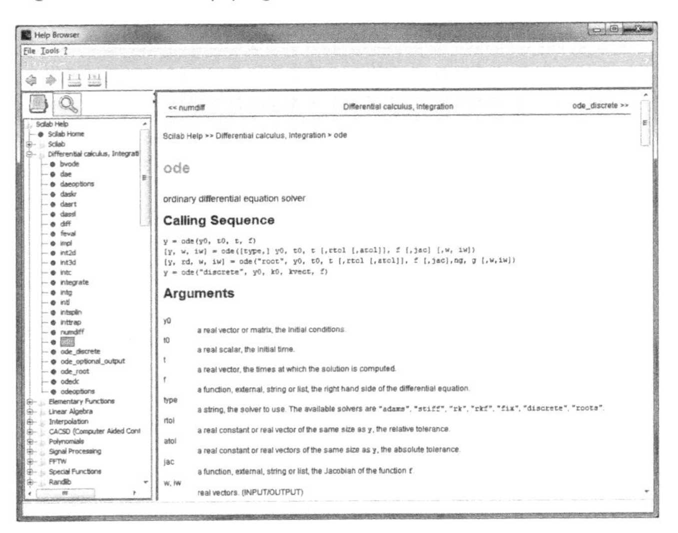

After reading this page, you will understand that to solve a differential equation, you need to define:

- the Scilab function `F(t,u)` which takes as argument a real number `t` (time) and a vector `u` (which, here, has twelve components)
- a vector `t` which contains all the values of time for which we need to compute the values of `u`
- a vector `u0` which contains the initial values of `u`

The only thing left to do is to apply the function `ode` with the format `u=ode(u0,0,t,F)` to retrieve the results inside a vector `u`. Be careful, in the results, line `i` of `u` corresponds to the values of the ith coordinate of `u` for each time of the vector `t` (starting at time 0). These are the bases of the animation shown here. The `ode` command lets you compute the positions without writing all the calculations of Euler's method (as we had done for the case of the pendulum hanging from a spring which was covered in Section 24.1, *The spring pendulum*), which greatly simplifies the code.

Tip › *In addition to the low-level features explained in the book, Scilab has numerous high-level functions capable of solving complex but recurrent problems such as* **ode**. *Don't hesitate to look for a feature inside the help documentation before starting to write a function that may already exist.*

We can now move on to the explanation of the script that corresponds to this simulation:

```
//*****************************************************
// simulation of the perturbed trajectory of a comet
//*****************************************************
//  parameterization of a sphere
function [x,y,z]=sphere(theta,phi)
   A=0.1,B=0.01
   x=A*cos(phi).*cos(theta)
   y=A*cos(phi).*sin(theta)
   z=B*sin(phi)
endfunction
// function to draw a sphere
function plot_sphere(x,y,z)
   phi=[0:0.1:2*3.15];
   theta=[2*3.15:-0.05:0];
   [dx,dy,dz]=eval3dp(sphere,theta,phi);
   surf(x+dx,y+dy,z+dz);
endfunction
// function to plot the z=0 plane
function plot_ecliptic(ebox)
   x=[ebox(1);ebox(2)]
   y=[ebox(3);ebox(4)]
   z=zeros(2,2)
   surf(x,y,z)
endfunction

// functions calculating the gravitational forces
function [u2]=force_g(t,u,mass)
   module=-G*mass*((u(1)^2+u(2)^2+u(3)^2)^(-3/2))
   u2=[module*u(1); module*u(2); module*u(3)]
endfunction

function [du]=force(t,u,mass0,mass1)
   u1=[u(1);u(2);u(3)]
   du1=[u(4);u(5);u(6)]
   u2=[u(7);u(8);u(9)]
   du2=[u(10);u(11);u(12)]
   ddu1=force_g(t,u1,mass0)
   ddu2=force_g(t,u2,mass0)+force_g(t,u2-u1,mass1)
   du=[du1;ddu1;du2;ddu2]
endfunction

// constants
G=0.04;
m0=1000;
m1=1;
dt=0.05;
T=50;
dx=0.5;
dy=0.5;
```

```
dz=0.5;
alpha=65;
Beta=150;

//**********************************************************
// trajectory calculations
//**********************************************************
// initial coordinates of the planet
x1=5;y1=0;z1=0;vx1=0;vy1=2.5;vz1=0;
// initial coordinates of the comet
x2=6;y2=6;z2=0.21;vx2=-2;vy2=-0.5;vz2=-0.1;
//solve the differential equation using ode
t=[0:dt:T];
u0=[x1; y1; z1; vx1; vy1; vz1; x2; y2; z2; vx2; vy2; vz2];
u=ode(u0,0,t,list(force,m0,m1));
// retrieve results
X=[u(1,:)',u(7,:)'];
Y=[u(2,:)',u(8,:)'];
Z=[u(3,:)',u(9,:)'];

//**********************************************************
// launch the graphics window
//**********************************************************
ebox=[min(X),max(X),min(Y),max(Y),min(Z),max(Z)];
N=length(t);                        // number of steps
drawlater()
plot_ecliptic(ebox) // plot the ecliptic plane
plot_sphere(0,0,0)                  // sun
plot_sphere(X(1,1),Y(1,1),Z(1,1))  // planet
plot_sphere(X(1,2),Y(1,2),Z(1,2))  // comet
A=gca();
A.axes_visible=["off" "off" "off"];
A.rotation_angles=[alpha Beta];
A.data_bounds=ebox;
drawnow()

//**********************************************************
// main loop creates the graphical animation
//**********************************************************
for k=1:5:N
   Beta=Beta+k/300;                          //  view angle
   realtimeinit(0.05)                       // unit of time
   drawlater()                // open the graphical buffer
   clf()                     // erase the graphical buffer
   plot_ecliptic(ebox)        // plot on ecliptic plane
   param3d1(X(1:k,:),Y(1:k,:),...            // display the
   list(Z(1:k,:),[5,2]))                    // trajectories
   plot_sphere(0,0,0)                            // the sun
   plot_sphere(X(k,1),Y(k,1),Z(k,1))          // the planet
   plot_sphere(X(k,2),Y(k,2),Z(k,2))           // the comet
   title('comet dynamics simulation : t='+msprintf(...
   '%2.2f',t(k))+'/'+string(T)+' years')           // title
   xinfo(string(t(k)))                      // display time
   A=gca();                  // resize the graphics window
   A.axes_visible=["off" "off" "off"];
   A.rotation_angles=[alpha Beta];     // rotate pt of vue
```

```
    A.data_bounds=ebox;
    drawnow()                  // display graphical buffer
    realtime(k)              // pause to adjust display rate
end
```

As with the pendulum hanging from a spring, we can split the script up into several sections.

1. The script starts by defining the three functions which will be used for displays:
   - `sphere` computes the parametric surface that corresponds to the spheres which symbolize the star, planet and comet.
   - `plot_sphere` calls the functions `eval3dp` and `surf` in order to draw the sphere defined in the previous function, centered at the point of coordinates `(x,y,z)` (provided as argument).
   - `plot_ecliptic` uses the `surf` function to draw the `z=0` plane over the whole area of the graphics window. To simplify, it takes as argument the matrix `ebox` that corresponds to the `data_bounds` property of the current Axes handle.

```
// parameterization of a sphere
function [x,y,z]=sphere(theta,phi)
   A=0.1,B=0.01
   x=A*cos(phi).*cos(theta)
   y=A*cos(phi).*sin(theta)
   z=B*sin(phi)
endfunction
// function to draw a sphere
function plot_sphere(x,y,z)
   phi=[0:0.1:2*3.15];
   theta=[2*3.15:-0.05:0];
   [dx,dy,dz]=eval3dp(sphere,theta,phi);
   surf(x+dx,y+dy,z+dz);
endfunction
// function to plot the z=0 plane
function plot_ecliptic(ebox)
   x=[ebox(1);ebox(2)]
   y=[ebox(3);ebox(4)]
   z=zeros(2,2)
   surf(x,y,z)
endfunction
```

2. We define the two other functions that are used to compute the gravitational forces that appear inside the equations:
   - The first function `force_g` computes the gravitational force between two objects.
   - The second function `force` gets passed as argument to `ode` to compute the solution to the three-body problem.

```
// functions calculating the gravitational forces
function [u2]=force_g(t,u,mass)
   module=-G*mass*((u(1)^2+u(2)^2+u(3)^2)^(-3/2))
   u2=[module*u(1); module*u(2); module*u(3)]
endfunction

function [du]=force(t,u,mass0,mass1)
   u1=[u(1);u(2);u(3)]
   du1=[u(4);u(5);u(6)]
   u2=[u(7);u(8);u(9)]
   du2=[u(10);u(11);u(12)]
   ddu1=force_g(t,u1,mass0)
   ddu2=force_g(t,u2,mass0)+force_g(t,u2-u1,mass1)
   du=[du1;ddu1;du2;ddu2]
endfunction
```

To improve readability, the `force` function calls the `force_g` function. We have also added more parameters including the mass of the star and planet that are used in the equations: this makes them easier to modify if needed.

3. We define all the constants necessary for the calculations:
   - the gravitational constant `G`
   - the star mass, `m0`, and the planet mass, `m1`
   - the time step `dt` and the maximum simulation time `T`
   - the dimensions `dx, dy, dz` of the sphere used to draw the star, planet and comet
   - the view andles `Alpha, Beta` to display the scene in three dimensions

```
// constants
G=0.04;
m0=1000;
m1=1;
dt=0.05;
T=50;
dx=0.5;
dy=0.5;
dz=0.5;
alpha=65;
Beta=150;
```

4. We perform the trajectory computation with `ode`:
   - The initial data are gathered inside the vector `u0`.
   - The solution's calculation is performed by the line `u=ode(u0,0,t,list(force,m0,m1));`.

- Finally, we rewrite the coordinates of the trajectory's points for the planet and comet inside three matrices `X, Y, Z`, which simplifies the process of displaying the trajectories later on.

Inside the three matrices `X, Y, Z`, the first column corresponds to the planet's data and the second to the comet's.

```
// initial coordinates of the planet
x1=5;y1=0;z1=0;vx1=0;vy1=2.5;vz1=0;
// initial coordinates of the comet
x2=6;y2=6;z2=0.21;vx2=-2;vy2=-0.5;vz2=-0.1;
// solve the differential equation using ode
t=[0:dt:T];
u0=[x1; y1; z1; vx1; vy1; vz1; x2; y2; z2; vx2; vy2; vz2];
u=ode(u0,0,t,list(force,m0,m1));
// retrieve results
X=[u(1,:)',u(7,:)'];
Y=[u(2,:)',u(8,:)'];
Z=[u(3,:)',u(9,:)'];
```

5. We are then ready to launch the display inside a graphics window:
   - The dimensions of the graphics window are computed inside `ebox` as a function of the extremal values of the object positions in `X, Y, Z`.
   - The different elements are then plotted with the appropriate functions: `sphere`, `plot_sphere`, `plot_ecliptic`.

```
ebox=[min(X),max(X),min(Y),max(Y),min(Z),max(Z)];
N=length(t);                          // number of steps
drawlater()
plot_ecliptic(ebox) // plot the ecliptic plane
plot_sphere(0,0,0)                    // sun
plot_sphere(X(1,1),Y(1,1),Z(1,1))  // planet
plot_sphere(X(1,2),Y(1,2),Z(1,2))  // comet
A=gca();
A.axes_visible=["off" "off" "off"];
A.rotation_angles=[alpha Beta];
A.data_bounds=ebox;
drawnow()
```

6. Finally, the `for` loop takes care of displaying the animation:
   - The trajectories are drawn with `param3d1` and the other elements are displayed using the functions `sphere`, `plot_sphere`, `plot_ecliptic`.
   - We change the values of the current Axes handles, specifically `data_bounds`, to rotate the observation point during the simulation.

- We use the properties of the graphics buffer, along with `drawlater()` and `drawnow()`, and also `sleep(30)` to achieve a smoother display.

```
for k=1:5:N
   Beta=Beta+k/300;                                    //  view angle
   realtimeinit(0.05)                                  // unit of time
   drawlater()                          // open the graphical buffer
   clf()                               // erase the graphical buffer
   plot_ecliptic(ebox)                  // plot on ecliptic plane
   param3d1(X(1:k,:),Y(1:k,:),...                       // display the
   list(Z(1:k,:),[5,2]))                               // trajectories
   plot_sphere(0,0,0)                                       // the sun
   plot_sphere(X(k,1),Y(k,1),Z(k,1))                     // the planet
   plot_sphere(X(k,2),Y(k,2),Z(k,2))                      // the comet
   title('comet dynamics simulation : t='+msprintf(...
   '%2.2f',t(k))+'/'+string(T)+' years')                      // title
   xinfo(string(t(k)))                                 // display time
   A=gca();                            // resize the graphics window
   A.axes_visible=["off" "off" "off"];
   A.rotation_angles=[alpha Beta];          // rotate pt of vue
   A.data_bounds=ebox;
   drawnow()                            // display graphical buffer
   realtime(k)                      // pause to adjust display rate
end
```

Index (commands excluded)

Symbols

A

B

C

D

Commands

Symbols

A

B

C

D

E

F

G

H

I

J

K

L

Q

R

S

Y

Z

About the authors

Philippe Roux

Philippe Roux is Associate Professor of Mathematics at the Institut universitaire de technologie de Lannion and holds a PhD in Mathematics. He uses Scilab to teach mathematics and its applications to students pursuing technology degrees in Computer Science. His experience teaching since 2001 has served as a basis for this introduction to Scilab.

You can find him online at his personal website http://perso.univ-rennes1.fr/philippe.roux/ or send him an email at philippe.roux@univ-rennes1.fr.

translated by Perrine Mathieu

Perrine Mathieu is a Franco-American aerospace engineer with a Bachelor of Science in Physics from McGill University and a Master of Science in Aerospace Engineering from the University of Texas at Austin. She has worked with Scilab through her profession and studies on both side of the Atlantic. You can get in touch with her at perrine.mathieu@utexas.edu

Lightning Source UK Ltd.
Milton Keynes UK
UKOW06f0829100516

273942UK00017B/416/P